1st International Conference on Engineering for Sustainable World (ICESW 2017)

AA006339

IOP Conference Series: Materials Science and Engineering
Volume 391

Ota, Nigeria
3 – 7 July 2017

ISBN: 978-1-5108-6786-4
ISSN: 1757-8981

Printed from e-media with permission by:

Curran Associates, Inc.
57 Morehouse Lane
Red Hook, NY 12571

proceedings.com

Some format issues inherent in the e-media version may also appear in this print version.

Copyright© (2017) by the Institute of Physics
All rights reserved. The material featured in this book is subject to
IOP copyright protection, unless otherwise indicated.

Printed by Curran Associates, Inc. (2018)

For permission requests, please contact the Institute of Physics
at the address below.

Institute of Physics
Dirac House, Temple Back
Bristol BS1 6BE UK

Phone: 44 1 17 929 7481
Fax: 44 1 17 920 0979

techtracking@iop.org

Additional copies of this publication are available from:

Curran Associates, Inc.
57 Morehouse Lane
Red Hook, NY 12571 USA
Phone: 845-758-0400
Fax: 845-758-2633
Email: curran@proceedings.com
Web: www.proceedings.com

Table of contents

Volume 391

The 1st International Conference on Engineering for Sustainable World (ICESW)

3–7 July 2017, Ota, Nigeria

Accepted papers received: 6 July 2018
Published online: 23 July 2018

Preface

Papers

IOP Conf. Series: Materials Science and Engineering **391** (2018) 011001 doi:10.1088/1757-899X/391/1/011001

ICESW2017

PREFACE

This is a proceeding of the 1st International Conference on Engineering for Sustainable World (ICESW).

ICESW brings together international experts, scholars, policy makers, industrialists, etc to discuss contemporary and cutting-edge research in engineering management, innovation, energy development and clean technology. It also provides a golden opportunity to meet experts and develop new collaborations on the fundamentals, applications and products from various fields such as: sustainable energy, sustainable cities, engineering innovations, clean technology/production, sustainable materials and engineering, mechatronics, renewable energy systems etc. The conference strives to foster new collaborative and interdisciplinary relationships among emerging and established engineers, scientists, and policy makers.

ICESW' topics are built on the fundamentals of sustainability, with vigorous attention to all forms of engineering professions and enabling research tools including the following topics.

- Renewable Energy and Sustainability
- Engineering Innovations
- Sustainable Manufacturing Systems and Techniques
- Modelling and Simulations
- Clean Technology
- Local Contents in Manufacturing
- Chemistry and Physics of Material
- Sustainable Materials and Processing
- Engineering Policies and Education for Sustainable Development
- Mechatronics
- Thermal Engineering
- Sustainable Buildings and Cities
- Sustainable Product Design and Development
- Eco-Industrial Parks
- Industrial and Production Engineering

ORGANISING COMMITTEE

Nigeria:

- Israel Dunmade (Conference Chair), Covenant University ,Nigeria
- Sunday Oyedepo (Co-Chair), Covenant University, Nigeria
- Mfon Udo (Secretary) Covenant University, Nigeria
- Cleophas Loto, Covenant University, Nigeria
- Christian Bolu, Covenant University, Nigeria
- Festus Oyawale, Covenant University, Nigeria
- Anthony Inegbenebor, Covenant University, Nigeria

Content from this work may be used under the terms of the Creative Commons Attribution 3.0 licence. Any further distribution of this work must maintain attribution to the author(s) and the title of the work, journal citation and DOI.
Published under licence by IOP Publishing Ltd

- Mufutau Waheed, Federal University of Agriculture, Nigeria
- David Esezobor, University of Lagos, Nigeria
- Samuel Onawumi, LAUTECH, Ogbomosho, Nigeria
- Richard Fagbenle – University of Ibadan, Nigeria
- Kayode Simonyan, Michael Okpara University of Agriculture, Nigeria
- Oluseyi Ajayi, Covenant University, Nigeria
- Philip Babalola, Covenant University, Nigeria
- Yinka Ohunakin, Covenant University, Nigeria
- Oluseun Kilanko, Covenant University, Nigeria
- Joshua O. Okeniyi, Covenant University, Nigeria
- Sunday Fayomi, Covenant University, Nigeria
- Sociis Okolie, Covenant University, Nigeria
- Tajudeen Salau, University of Ibadan, Nigeria
- Echana Oyetunji, University of Ibadan, Nigeria
- Wahabi Asiru, Federal Institute of Industrial Research, Nigeria
- Samuel Adejuyigbe, Federal University, Oye-Ekiti, Nigeria
- Lasode Muyiwa, University of Ilorin, Nigeria
- John Olorunmaiye, University of Ilorin, Nigeria
- Stephen Adavbiele, Ambrose Ali University, Ekpoma, Nigeria
- Tunde Damisa, Federal University of Petroleum, Nigeria
- Olarewaju Oyewola, University of Ibadan, Nigeria

International:

- Bert Bras, Georgia Institute of Technology, Atlanta, USA
- Hart Martin, Tomas Bata University, Zlin, Czech Republic
- Michael Packlanather, Cardiff University, Wales, UK
- Giampaolo Campana, University of Bologna, Italy
- Dele Odeshi, University of Saskatchewan, Saakatoon, Canada
- Joshua Yusuf, All Saint's University of Medicine, St. Vincent & Grenadines
- Oladiran Fashina, Auburn University, Alabama, USA
- Michael Ngadi, McGill University, Montreal, Canada
- Klein Ileleji, Purdue University, Indiana, USA
- Luc Hebou, University of Maryland, USA
- Tunde Bello-Ochende, Cape Town University, South Africa
- Muyiwa Adaramola, University of Norway

IOP Conf. Series: Materials Science and Engineering **391** (2018) 011002 doi:10.1088/1757-899X/391/1/011002

Peer review statement

All papers published in this volume of *IOP Conference Series: Materials Science and Engineering* have been peer reviewed through processes administered by the proceedings Editors. Reviews were conducted by expert referees to the professional and scientific standards expected of a proceedings journal published by IOP Publishing.

Content from this work may be used under the terms of the Creative Commons Attribution 3.0 licence. Any further distribution of this work must maintain attribution to the author(s) and the title of the work, journal citation and DOI.
Published under licence by IOP Publishing Ltd

Investigation of the heat transfer effect of Ni/R134a nanorefrigerant in a mobile hybrid powered vapour compression refrigerator

[1]Oluseyi O. Ajayi, [1]Oghenefegor O. Useh, [1]Solomon O. Banjo, [2]Felicia T. Oweoye, [1]Ameh Attabo, [1]Mercy Ogbonnaya, [1]Imhade P. Okokpujie, [1]Enesi, Y. Salawu

[1]Mechanocal Engineering Department, Covenant University, P.M.B. 1023, Ota, Ogun State, Nigeria
[2]Department of Industrial Chemistry, Covenant University, P.M.B. 1023, Ota, Ogun State, Nigeria
*Correspondence: oluseyi.ajayi@covenantuniversity.edu.ng

Abstract. The study investigated the effect of the use of 0.04%Ni/R134a nanorefrigerant as working fluid in a vapour compression refrigeration system designed to work with hybrid power source of grid-electricity/battery/solar power. The nanoparticle was prepared via the one step method and dispersed into the mineral oil. R134a was employed as the base refrigerant and the performance of the system with the nanorefrigerant was compared, in terms of the coefficient of performance, pull down time and power consumption, with that of the base refrigerant. The results showed that the refrigeration system performed better with the nanorefrigerant as working fluid with any of the power sources and also with the hybrid solar/battery power. The system with nanorefrigerant performed better with improved coefficient of performance in the range of between 7.05% when used with solar/battery as power source and 14.12% when with only battery as power source. The system's pull-down time showed far better results with the nanorefrigerant compared with base refrigerant. However, the system with nanorefrigerant slightly consumed more power with the highest consumption being 28.7% (with battery only as power source) higher that that with the base refrigerant. Based on the outcome of the results, the system can be deployed for vaccine preservation in off-grid remote arears, and also for domestic and commercial applications.
Key words: Nanorefrigerant, Energy System, Hybrid Power System, Solar Energy, Off-grid system, vaccine preservation

1. Introduction

Nanotechnology is a section of science and technology regarding the modification and use of particles in the atomic and molecular order (nanoparticles). A form of this technology is seen in nanofluids. Nanofluids are simply fluids that have a composition of nanoparticles and base fluid. According to application, nanofluids are classified as heat transfer nanofluids, tribological nanofluids, surfactant and coating nanofluids, chemical nanofluids, process/extraction nanofluids, environmental (pollution cleaning) nanofluids, bio- and pharmaceutical nanofluid and medical nanofluids (drug delivery, functional and tissue-cell interaction) [1]. The focus of this study is on that which belonged to the class of heat transfer nanofluids, in this case nanorefrigerant. Nanorefrigerants are special type of nanoluids which comprises of the mixture of nanoparticles and base working fluids. According to [2], there are three main advantages derived from the application of nanorefrigerants. These advantages are related to its ability to enhance the solubility between the lubricant and the refrigerant, the ability to influence thermal conductivity and heat transfer characteristics of the refrigerant and also the ability to decrease the friction coefficient and wear rate of the lubricant.

Content from this work may be used under the terms of the Creative Commons Attribution 3.0 licence. Any further distribution of this work must maintain attribution to the author(s) and the title of the work, journal citation and DOI.
Published under licence by IOP Publishing Ltd

Some studies have been conducted on the effect of nanoparticles in the working fluid of vapour compression refrigeration system. Such studies have focused majorly on the performance of refrigeration system based on its freezing capacity, coefficient of performance and power consumption. For instance, [3], investigated experimentally the reliability and performance of a domestic refrigerator with nanoparticles dispersed in the working fluid. TiO_2 nanoparticles were mixed with mineral oil and used as a lubricant in a refrigerator system using 1,1,1,2-tetrafluoroethane (HFC134a) refrigerant. The performance of the refrigerator was investigated using energy consumption tests and freeze capacity. From the experiments, the results showed that HFC134a and mineral oil with TiO_2 nanoparticles was environmentally safe and worked normally. Also, the refrigerator performance was better than the conventional HFC134a and POE (Poly ol-ester) oil system, and the energy consumption was 26.1% lower when 0.1% mass fraction of TiO_2 nanoparticles was employed. [4] numerically investigated the heat transfer enhancement of a domestic refrigerator operated with Al_2O_3 nanorefrigerant. The study investigated the impact of different nanofluids on the heat transfer effectiveness of the system. Three base fluids were used for the experiment - pure POE oil, mineral oil and mineral oil + alumina nanoparticles (0.06% mass fraction) as lubricants for the compressor of the system. The results showed that the pull down time for the system with nanolubricant (Al_2O_3/mineral oil) was lowest. It was also observed that temperature drop in the condenser with nanorefrigerant was high, with the exit of the condenser being 39°C and subcooling obtained at 3°C. Further to this, it was found that the coefficient of performance (COP) increased by 19.6% and the power consumption reduced by 11.5% when nanorefrigerants were used.

In addition, [5] investigated the effect of dispersing a low concentration of TiO_2 nanoparticles in the mineral-oil lubricant on its viscosity and lubrication characteristics as well as on the overall performance of a Vapour Compression System using R12 (Dichlorodifluoromethane) as the working fluid. The outcome showed that there was an enhancement in the COP of the refrigeration system. It was also found that the addition of low volume fractions of titanium oxide nanoparticles in the mineral oil brought about a reduction in the compressor work. The results indicated that the increase in viscosity due to the addition of nanoparticles was more prominent in the lower temperature range. It was also realized that the friction coefficient reduces till the volume fraction got to 0.01% and then increased for volume fractions above that. Furthermore, the results demonstrated that with the presence of nanoparticles in the mineral oil, the average heat transfer rate increased by about 3.6%, compressor work decreased by about 1.1% and resulted in COP improvement of 17%. [6] investigated experimentally the reliability and performance of a vapour compression refrigeration system with ZnO nanorefrigerant. The base refrigerant used was R152a and, the particle size and volume concentrations were 50 nm and varying 0.1%, 0.3% and 0.5%, respectively. About 150 g of the refrigerant was charged into the system. The results showed that ZnO nanorefrigerant works normally and safely in the system. The performance of the system was significantly improved with 21% lower energy consumption when 0.5% volume concentration was used. Both the suction and discharge pressure were lowered by 10.5%. Also, the temperature of the evaporator reduced by 6%, and the results of the pull-down time and COP showed great improvement. This study therefore aims to extend the database of suitable nanorefrigerants that can be employed for the performance enhancement of vapour compression refrigeration systems. It employed nickel particle-dispersed into capella D oil and R134a base refrigerant to form the nanorefrigerant. However, unlike previous studies, this study used a hybrid grid-electricity/solar/battery powered vapour compression system designed for the purpose of the investigation. It investigated the performance, energy saving capacity and the compatibility of the vapour compression refrigeration system with solar power. The purpose was to have a system that can be suitably employed as off-grid refrigeration system for rural and commercial applications, and also for vaccine preservation.

2. Materials and Methods

It has been discovered that nanofluid can be prepared in two basic ways. They are, one-step method and two-step method. Popularly used for preparation is the two-step method where the nanoparticles are directly mixed with the fluid. This method is more economical but has some setbacks, like the aggregation and sedimentation of nanoparticles in the nanofluid, which impairs the performance of the

IOP Conf. Series: Materials Science and Engineering **391** (2018) 012001 doi:10.1088/1757-899X/391/1/012001

nanofluid. On the other hand, the one-step method has the nanoparticles synthesized from a salt into the base fluid. This preparation though more expensive than the two-step method, produces nanofluid without sedimentation of the nanoparticles. Nanofluid produced with the one-step method have been found to be more effective than the other. For purpose of this study, the nanofluid was prepared using the one-step method. Nickel nanoparticle prepared by the one step method was dispersed directly into Cappella D oil and the system was charged with R134a refrigerant. The suitability and performance of 0.04% Ni/R134a nanorefrigerant was investigated and compared with when the base refrigerant of R134a was used.

The refrigerating system used for this experiment was a hybrid refrigerator that can use solar/battery, only battery and grid electricity as power sources. The experiments were carried out with each of the power sources to determine the system's effectiveness. The refrigerator was made up of a cooler that acts as the evaporator, plate-fin condenser, a hermetic compressor, capillary tube, an inverter that converts the DC to AC, a solar panel, a charge controller and a 150 Ah Deep-Cycle battery. Figs. 1 to 3 displays the results obtained when grid-electricity was used as the power source, while Figs. 4 – 7 are those obtained when the power source was just the battery. Only battery as power source was carried out to determine the performance when there is no solar capacity. Moreover, Figs. 8 – 11 displays the results obtained when a hybrid solar/battery power system was employed for the refrigeration system.

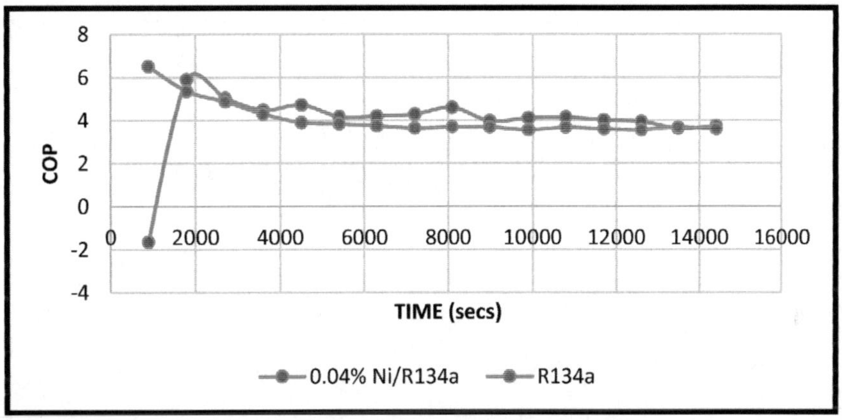

Fig. 1: Comparison of the coefficient of performance of the system with and without nanorefrigerant when grid electricity was the power source

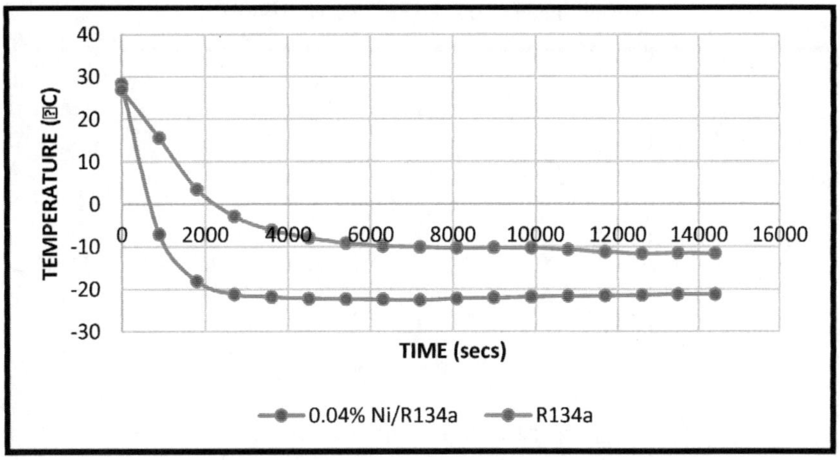

Fig. 2: Plot showing the pull down time of the system with and without nanorefrigerant when grid electricity was the power source

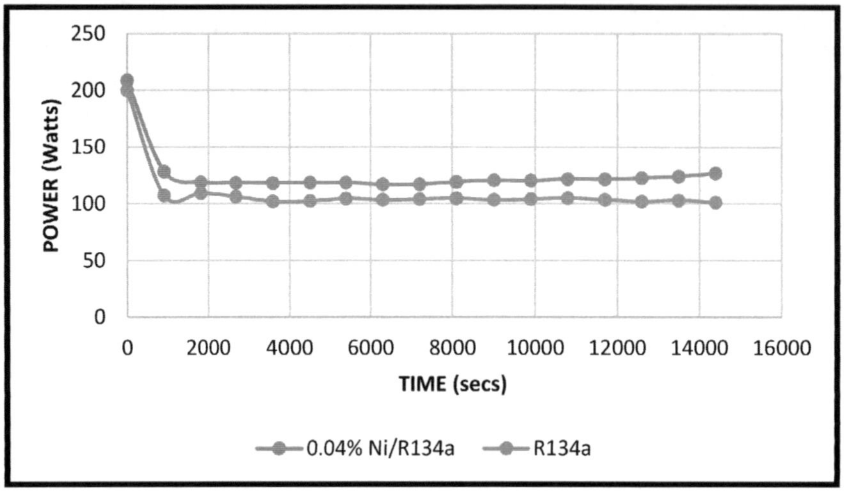

Fig. 3: Variation of power consumption of the system with and without nanorefrigerant when grid electricity was the power source

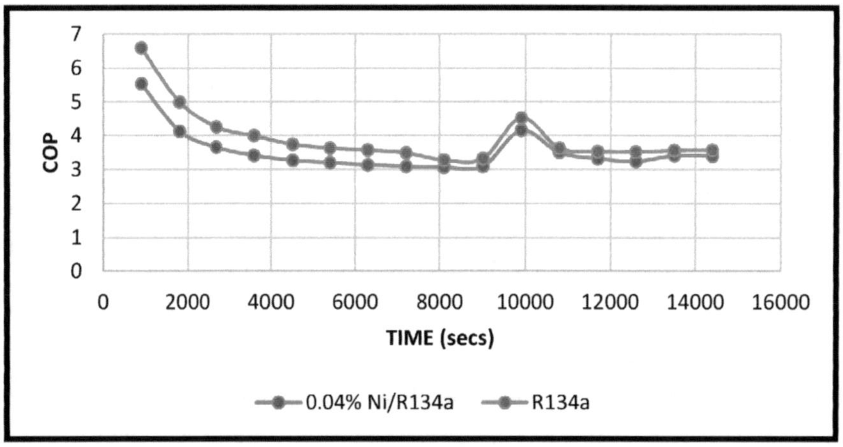

Fig. 4: Comparison of the coefficient of performance of the system with and without nanorefrigerant when battery was the only power source

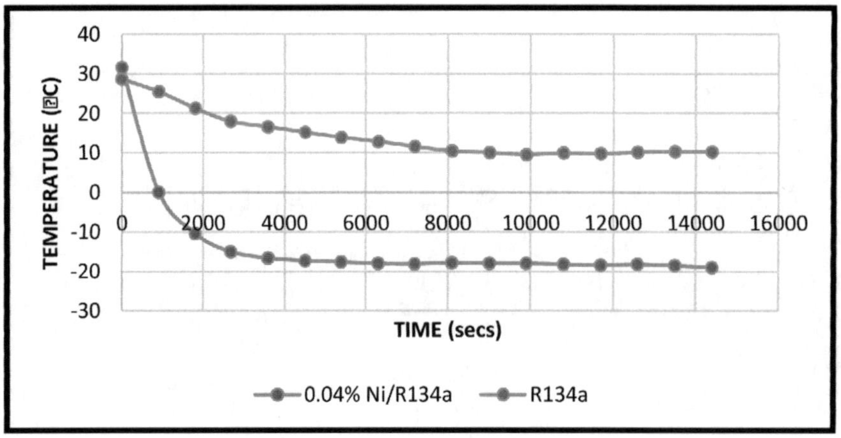

Fig. 5: Plot showing the pull down time of the system with and without nanorefrigerant when battery was the only power source

IOP Conf. Series: Materials Science and Engineering **391** (2018) 012001 doi:10.1088/1757-899X/391/1/012001

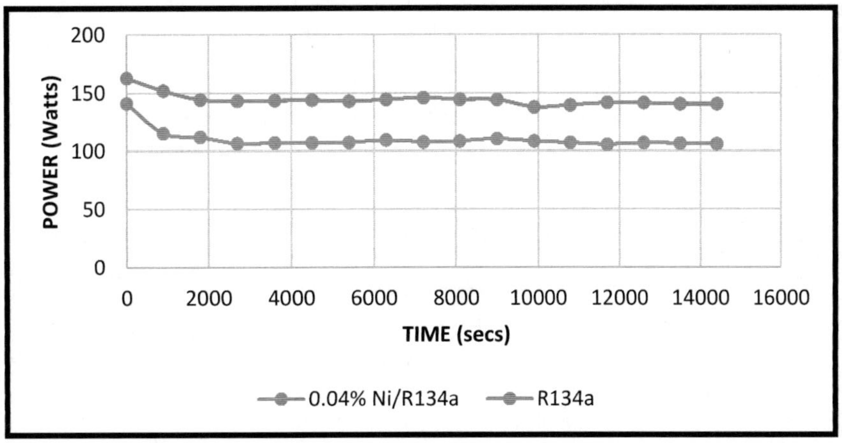

Fig. 6: Variation of power consumption of the system with and without nanorefrigerant when battery was the only power source

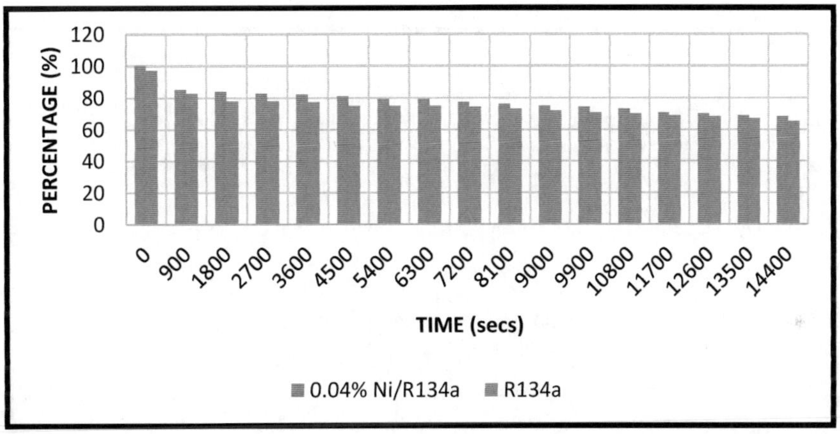

Fig. 7: Variation of battery discharge rate when battery was the only power source

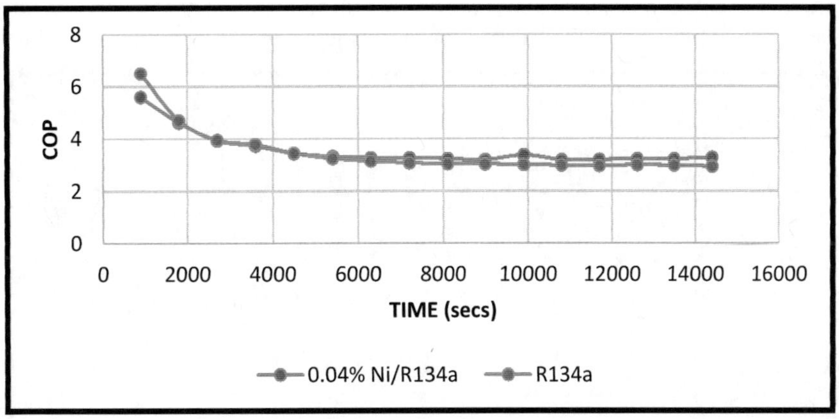

Fig. 8: Comparison of the coefficient of performance of the system with and without nanorefrigerant when hybrid solar/battery was the power source

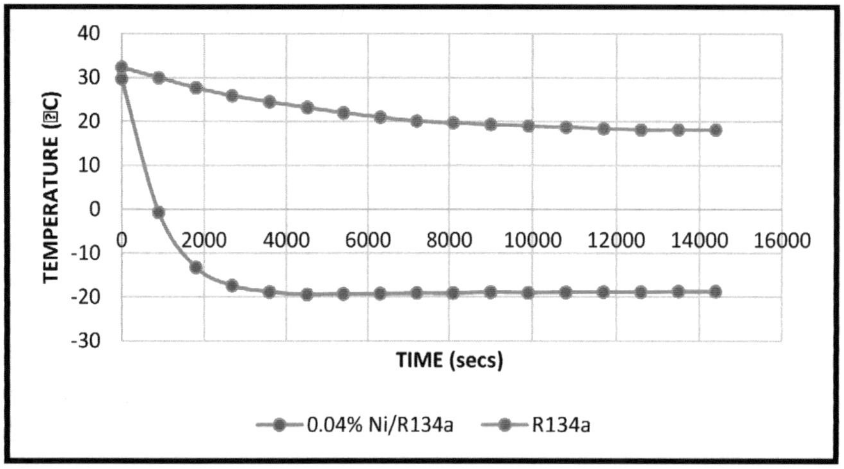

Fig. 9: Plot showing the pull down time of the system with and without nanorefrigerant when hybrid solar/battery was the power source

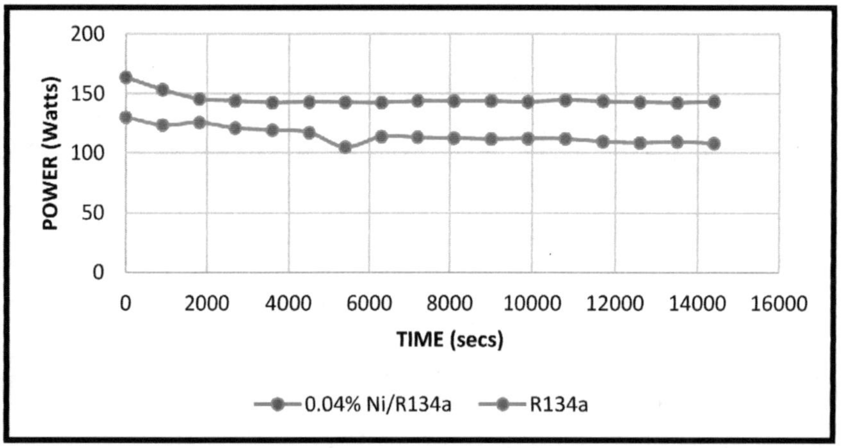

Fig. 10: Variation of power consumption of the system with and without nanorefrigerant when hybrid solar/battery was the power source

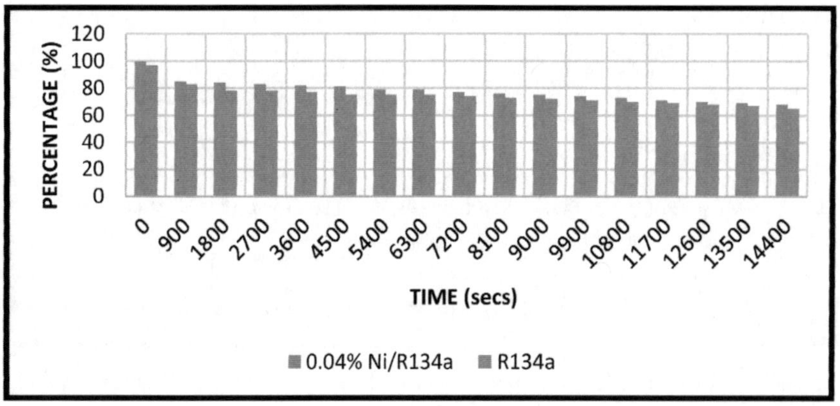

Fig. 11: Variation of battery discharge rate with hybrid solar/battery as the power source

3. Result and Discussion

Fig. 1 shows that the nanorefrigerant performed better than the conventional R134a when grid-electricity was employed as the power source with a better COP value of 12.8% higher. In the same vein, the pull-down time values (Fig. 2) showed better performance with the 0.04% Ni/R134a. The difference in temperature attained for the nanorefrigerant system was about 82.8% lower than that of R134a. The time taken to attain the lowest temperature stability was in the neighborhood of between 30 to 45 minutes. The magnitude of power consumption when grid-electricity was employed is shown in Fig. 3. The figure shows that, the system with nanorefrigerant slightly consumed a higher magnitude of power with the difference being about 17.4% higher. This result is however a contradiction when compared with other results of similar tests [7]; [8], pointing to a need for further investigation.

Moreover, when only battery was used as the power source, the COP results showed that the nanorefrigerant did not performed as the conventional R134a as shown in Fig. 4. The reason may be due to the fact that, the initial power requirement from the battery to start the compressor and make it function was somewhat inadequate when employed on the nanorefrigerated system. However, after operation for over two hours, the performance became very close. Thus, the overall COP drop from that of the conventional R134a was about 14.12%. Despite this result, the system with 0.04% Ni/R134a operated with better pull-down time (Fig. 5), attaining twice lower temperatures within the refrigerated space. Worth noting is also the fact that, while the conventional refrigerant R134a was not able to take the temperature of the refrigerated space below 0°C, the 0.04% Ni/R134a took it up to – 18°C within the same time interval. The power consumption tests of Fig. 6 shows that the system consumed about 28.9% more power with nanorefrigerant as the working fluid, when compared to the conventional refrigerant. These results were supported by the COP values of Fig. 4. An investigation into the battery capacity showed that the rate of battery discharge was slower with nanorefrigerant as the working fluid when compared with the conventional refrigerant. The results are displayed in Fig. 7.

To further determine the suitability of the hybrid system for off-grid application, the system was powered with solar/battery arrangement. The performance (Fig. 8) showed that the system performed better with 0.04% Ni/R134a nanorefrigerant as against the conventional R134a. The COP value was higher with about 7.05%. The pull-down time results (Fig. 9) shows that the temperature attained in the refrigerated space, when the nanorefrigerant was used was more than twice lower than that of the conventional as was the case with only battery. The lowest temperature of the refrigerated space was as low as -19°C. In terms of the power consumption (Fig. 10), the refrigerator consumed power of about 15.9% more with nanorefrigerant as the working fluid, when compared to the conventional refrigerant. Unlike when only battery was used as the power source (Fig.7), Fig. 11 shows that as a result of the solar power, the battery was sustained at a high percentage. Also, the battery capacity in percentage, reduced at a slower rate with the nanorefrigerant as the working fluid when compared with the conventional refrigerant. Comparing Fig. 7 with Fig. 11 shows that very similar results were obtained for the two test scenarios.

4. Conclusion

The study investigated the suitability and applicability of employing 0.04% Ni/R134a with vapour compression refrigeration systems. It also demonstrated the potential of employing such nanorefrigerated system for an off-grid application, and also for vaccine preservation. The performance analyses showed that the nano-refined heat transfer fluid of 0.04% Ni/R134a is suitable as a working fluid for a vapour compression refrigeration system. The outcome also showed that the nanorefrigerant can be used in an off-grid vapour compression refrigerating system without any retrofitting. Going by the ability of the 0.04% Ni/R134a nanorefrigerant to drop the temperature of the refrigerating space to below 0°C within very short time, employing adequate thermostat with the refrigerating system will ensure the system's suitability for off-grid vaccine preservation. However, despite the good performance results, it was discovered that the 0.04% Ni/R134a nanorefrigerant consumed slightly more power than the conventional R134a refrigerating system. The result contradicts some of those of previously studies, thereby bringing about the need for further studies.

References

[1] Ajayi, O.O., Ibia, D.E., Ogbonnaya, M., Attabo, A., & Agarana, M. (2017). CFD analysis of nanorefrigerant through adiabatic capillary tube of vapour compression refrigeration system. *Procedia Manufacturing,* 7, 688 – 695

[2] lawi, O. A., Sidik, N. A. C., & Beriache, M. (2015). Applications of nanorefrigerant and nanolubricants in refrigeration, air-conditioning and heat pump systems: A review. *International Communications in Heat and Mass Transfer*, *68*, 91–97.

[3] Alawi, O. A., Sidik, N. A. C., & Kherbeet, A. S. (2015). Nanorefrigerant effects in heat transfer performance and energy consumption reduction: A review. *International Communications in Heat and Mass Transfer*, *69*, 76–83.

[4] Alawi, O. A., Sidik, N. A. C., & Mohammed, H. A. (2014). A comprehensive review of fundamentals, preparation and applications of nanorefrigerants. *International Communications in Heat and Mass Transfer*, *54*, 81–95.

[5] Attabo, A.M. (2015). Thermodynamic assessment of the effects of synthetic and organic nano-fluid/-particles on a vapour compression refrigeration system, Masters Dissertation, Covenant University, Ota, Nigeria.

[6] Bi, S. shan, Shi, L., & Zhang, L. li. (2008). Application of nanoparticles in domestic refrigerators. *Applied Thermal Engineering*, *28*(14–15), 1834–1843.

[7] Krishna Sabareesh, R., Gobinath, N., Sajith, V., Das, S., & Sobhan, C. B. (2012). Application of TiO2 nanoparticles as a lubricant-additive for vapor compression refrigeration systems – An experimental investigation. *International Journal of Refrigeration*, *35*(7), 1989–1996.

[8] Kumar, R. R., Sridhar, K., & Narasimha, M. (2013). Heat transfer enhancement in domestic refrigerator using R600a/mineral oil/nano-Al2O3 as working fluid. *International Journal of Computational Engineering Research*, *3*, 42–50.

[9] Kumar, R., & Singh, J. (2016). Effect of ZnO nanoparticles in R290/R600a (50/50) based vapour compression refrigeration system added via lubricant oil on compressor suction and discharge characteristics. *Heat and Mass Transfer/Waerme- Und Stoffuebertragung*, 1–9. https://doi.org/10.1007/s00231-016-1921-3

IOP Conf. Series: Materials Science and Engineering **391** (2018) 012002 doi:10.1088/1757-899X/391/1/012002

The Effect of TiO_2 Particulate Reinforcement on the Microstructure and Mechanical Properties of Binary Nano-Composite on Low Carbon Steel

A. A. Daniyan[1*], L. E. Umoru[1], A. P.I. Popoola[2] and O. S. I. Fayomi[2,3]

[1]Department of Materials Science and Engineering, Obafemi Awolowo University Ile-Ife, Nigeria
[2]Department of Chemical, Metallurgical and Materials Engineering, Tshwane University of Technology, P.M.B. X680, Pretoria, South Africa.
[3]Department of Mechanical Engineering, Covenant University, PMB 1023 Ota Nigeria
*Corresponding author: daniyan03@gmail.com; Tel: (+2348053777484)

Abstract- In this Research, the microstructure and mechanical properties of Zn-TiO_2 nano-composite, co- deposited on low carbon steel substrate were fabricated by electrocodeposition route. The structural performance of the composite coating was evaluated by scanning electron microscope (SEM) equipped with energy dispersive spectrometer (EDS). The mechanical property was carried out using a Durascan hardness tester. The result showed that the influence of TiO_2 nano particles greatly enhanced the hardness behaviour. The increase in hardness is ascribed to the uniform homogeneity and dispersion stability of particles in the aqueous solution containing the nano composite.
Keywords: Microstructure, Nano- composite, electrocodeposition, homogeneity and dispersion

1. Introduction

Ever since the invention of galvanic coating technology, electrolytic metal coatings have been used in quite a lot of areas due to their exceptional properties [1–3]. Electrodeposited composite coatings are a surface engineering developed on the basis of electrolytic conducting coating [4-6]. Metallic zinc has being the largely used electrodeposited metal to shield steel against corrosion. Nevertheless, the corrosion resistance render by zinc is inadequate. Lately, major efforts to improve the wear, hardness and corrosion resistance of metallic zinc coating, can be traced towards its composites with different materials [11-18]. It is also identified, in particular, that the physical, mechanical and electrochemical properties of materials can be improved with the introduction of nano powders as the second phase particles within the material of interest [13-18]. Due to this improvement, composites and alloys such as Zn-Al_2O_3, Zn-WO_3, Zn-Al and Zn-Ni have tremendously gained a wider range of applications in both marine and manufacturing industries recently as a better alternative for regular zinc plating [19-25]. Even though these coatings offer some benefits over zinc, but usually they cannot cathodically shield steel substrates in different type of corrosive environments [22-25]. Zinc alloy or composite deposited through electrodeposition have greater corrosion stability as compared to thermally obtained Zn alloy [6-8]. lately, successful zinc-aluminum electrodeposited coating on low carbon steel were done with the help of aluminium tri-chloride on zinc-aluminium deposited on the rest potential of yellowish and non-chromates zinc- aluminium [9].
In this work comparison of the microstructural and hardness of coated Zn and Zn-TiO_2 coatings on low

Content from this work may be used under the terms of the Creative Commons Attribution 3.0 licence. Any further distribution of this work must maintain attribution to the author(s) and the title of the work, journal citation and DOI.
Published under licence by IOP Publishing Ltd

carbon steelwas studied for the purpose of materials selection.

2. Experimental Procedure

2.1 Preparation of substrate

The measurement of the low carbon steel (substrate) used was (45 x 40 x 20) mm^3 piece and zinc pieces of (85 x 45 x 5) mm^3 were prepared as anodes. The low carbon steels pecimens' chemical composition is displayed in Table 1. The cathode was the low carbon coupons and anode was the commercially obtained pure zinc (about 99.99%). The low carbon steels specimens were polished mechanically, degreased and rinsed with water as described [6 - 8].

Table 1: Chemical composition of the low carbon steel used

Element	% Composition	Element	% Composition	Element	% Composition
C	0.330	Mo	0.100	Ti	0.04
Si	0.870	Ni	0.030	V	0.01
Mn	0.430	Cu	0.110	W	0.01
P	<0.003	Al	0.030	B	<0.002
S	>0.160	Co	0.030	Sn	0.0014
Cr	0.100	Nb	0.010	Fe	99.61

2.2 Formation of deposited coating

The low carbon steel substrate previously prepared was actuated by dipping into 10% hydrochloric acid solution for few seconds, then rinsed in deionized water. Analar chemicals grade and deionized water were used to formulate the coating solution at ordinary temperature before the coating process. The preparations of bath were formulated a day earlier and stir constantly at the rate of 400 rpm with continuous heating at 70°C during the course of the coating process, so as achieve standardized solution. The bath configurations used for the diverse coating matrix is as follows KCl 30 g/l, $ZnCl_2$ 120 g/l, TiO_2 nano particles 20g/l, Cetylpridinium Chloride 0.5g/l, 2- Butyne 1,4 diol 0.5g/l and Thiourea 10g/l.

The select of the coatings variables is in accordance with the study of preceding work of some of our collaborators [9]. The arranged zinc electrodes were coupled to the d.c power supply at varying current ranging from 1.0 and 1.5 A for 1200s (20 min) as shown in Figure 1. The prepared design plan for the coating is displayed in Table 2. The electrodeposited samples were rinsed in water and thereafter air-dried. Then, the plated steels were afterwards sectioned for characterization.

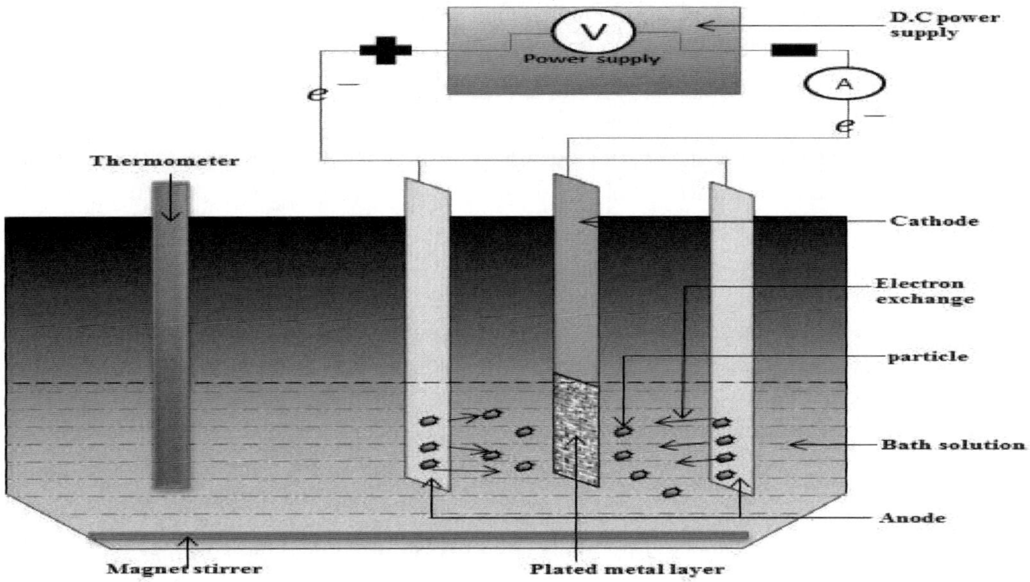

Figure 1: The diagram showing the schematic arrangement of the coating system (Fayomi and Popoola, 2014)

Table 2: Preparation of bath composition for Zn and Zn-TiO_2 nano-composite

Sample	Sample	Deposition time (min)	Current (A/m^2)
1	Zn-1.0A	20	560
2	Zn-1.5A	20	830
3	Zn-TiO_2-1.0A	20	560
4	Zn-TiO_2-1.5A	20	830

3. Coating Characterization

The obtained nano-composite coatings were characterized with the aids of JEOL FIELD EMMISSION (JSM – 7600F) Scanning electron microscope coupled with EDS. The structural arrangement was considered with XRD. Micro-hardness characterizations were obtain using a special Diamond pyramid indenter (EMCO) Test Dura-scan micro-hardness testers using a load of 10 g at a period of 20 s. The measured micro-hardness was carried out across the coated surface.

4. Results and Discussion

4.1 SEM/EDS studies of the deposited nanostructure composite

Figures 2 and 3 reveal the SEM/EDS image structure of Zn and Zn-TiO_2 fabricated at 1.5A (current density of 830 A/ m^2) respectively fabricated on low carbon steel. From the two Figures it is observable that the microstructural nodules of the coats were homogeneously spread on the surface. It is readily clear that, with the introduction of TiO_2 nanoparticles, a perceptible crystallite of the coating nodules along the edge was observed. Essentially there are two distinguishing phases, the first having a uniform steady deposits and the latter with agglomerate of reinforcements. The introduction of the TiO_2 in the zinc matrix could be traceable to the structure refined morphology coupled with good nodular particles, as can be seen in Figure 4 that presented the nodular Zn-TiO_2 particles ('Q' portion), in comparison to flaky ('P' portion) Zn particles. This is more pronounced at higher magnification (Figure 4). Figure 5 further revealed the morphology of the samples in nanometer range. The plated outlook and the coating edge of Zn-TiO_2 were quite amusing due to the fortification derived from the nano ceramics particulates of the composite in the metallic zinc matrix. The structural output was as anticipated since the nucleation development initiated from the metallic zinc load bearer, the particulates covering the nucleation points and reinforced the produced nanocomposite [10-11]. Additionally, from Figure 4, it is essential to remark that microstructural variation may be due to the incorporation TiO_2 nanoparticles which led to better deposition and improved orientation. Also, with reference to [11], the preparing factor with respect to the amount of additive can also influence the modification of the crystal route and surface value of a deposited surface. The penchant of electrodeposited materials to adhere and consequently form coatings with unique bright texture is influenced by the nature of the additive and the purity of the foundational particles [12 and 13]. The morphological orientations of the Zn matrix are clue of the strap up capability of the coating formed.

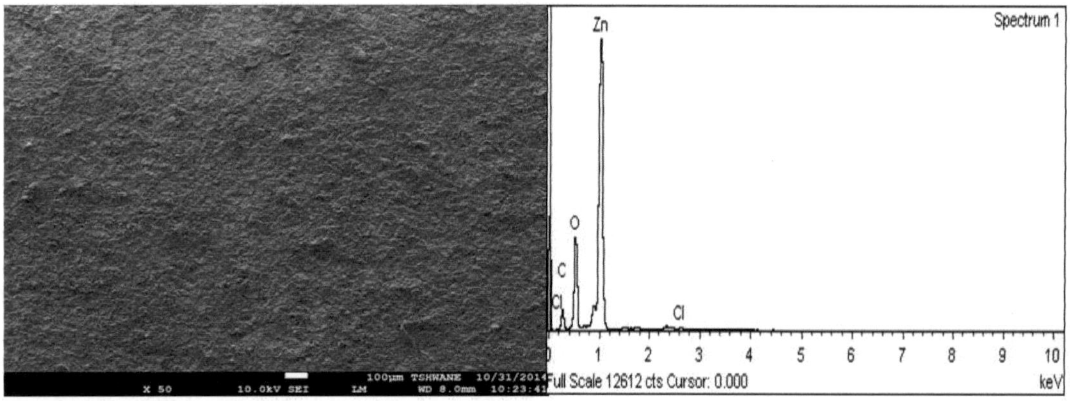

Figure 2: SEM/EDS of Zn-coated low carbon steelshowing the surface morphology

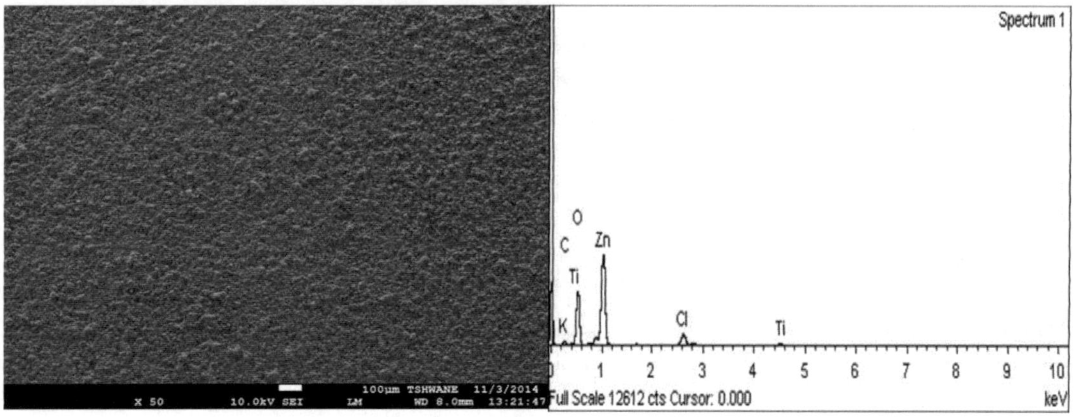

Figure 3: SEM/EDS of Zn-TiO_2 coated low carbon steel showing the surface morphology of the coatings

Figure 4: SEM/EDS of (a) Zn (b) Zn-TiO_2 coated low carbon steel showing the internal morphology of the coatings

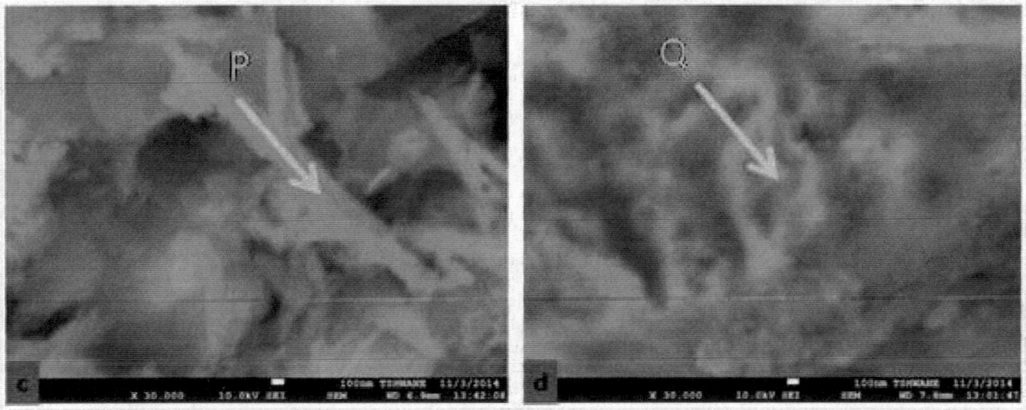

Figure 5: SEM/EDS of (a) Zn (b) Zn-TiO_2 coated low carbon steel in the nano- range showing the internal morphology of the coatings (P and Q are Zn and Zn-TiO_2-rich portions respectively)

4.2 Microhardness study

The microhardness study shown in Figure 6 was done so as to observe the effect of the composite microstructure and TiO_2 particle loading on the hardness behaviour of Zn-TiO_2 nanocomposite coatings.

In general, all the coatings formed, display excellent and unique improvement compared to the unplated low carbon steelsubstrate except Zn-1.0 A, which could be traced to the throwing power. However, the sample matrix Zn-TiO_2-1.5A showed the highest hardness property with the average value of 130.5 HVN. The outstanding enhancement of this matrix is primarily due to microstructural strengthening of the TiO_2 nanoparticles and the increased throwing power via current density (i.e., current density increased from 560 - 830 A/m^2). Figure 7 revealed the response of the composite coating to the heat treatment. The thermal deformations were observed in 250°C for 5hrs and mechanical responses of the coated samples were studied. The mechanical behaviour due to thermal treatment and porosity within the alloy surface could results in hardness distribution. Some literatures also agree to the fact that compression stress could significantly perk up the micro-hardness when it is much less than the ultimate strength of the coating but [14] revealed that the flaws such as porosity, macro-particle at the alloy surface will have unfavourable effect on the mechanical properties. The Figure 7 also compares the hardness properties of the composite coatings before and after thermal treatment. It should be noted that all the samples species increase in hardness after the thermal treatment, which is one of unique property of Nanoparticulate. The average micro hardness values for all the samples calculated indicates that, Zn-TiO_2- 1.5A nanocomposite coatings possessed the highest hardness value of 155.5 HVN after the thermal treatment. It can also be deduced from Figure 7 that, the hardness value of Zn after heat treatment increases from 114 to 143 HVN and 121 to 154 HVN when the current increases from 1.0 A to 1.5 A and that of Zn-TiO_2 increases from 129 to 145 HVN and 130.5 to 155.5 HVN as the current increased from 1.0A to 1.5 A. The coatings were excellent since the hardness values after heating increase tremendously in their values after thermal treatment [15]. The phenomenon of dispersion strengthening was also observed. The mechanisms of such strengthening are the grain refinement strengthening and the dispersion strengthening of TiO_2 deposited on the sample and the deposition current [16]. Hence, the microstructural improvement as a result of buildup of fine particulate and this build up enhances the hardness of the coating.

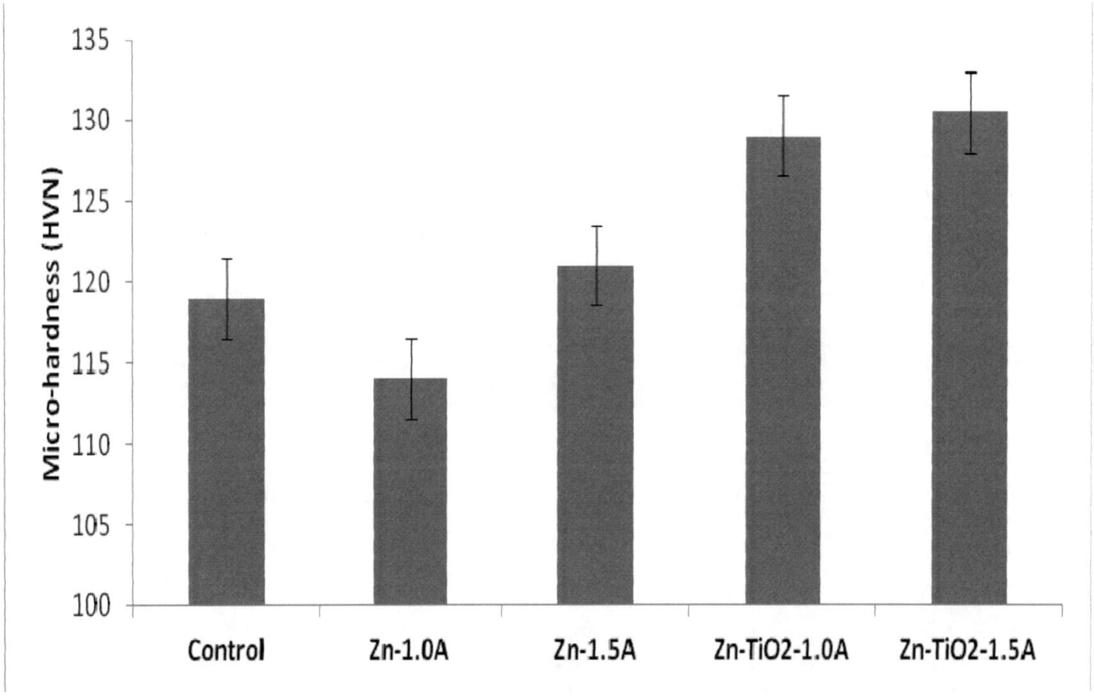

Figure 6: Hardness properties of varied current of Zn-TiO_2 nano-composite deposited on low carbon steel before heat treatment

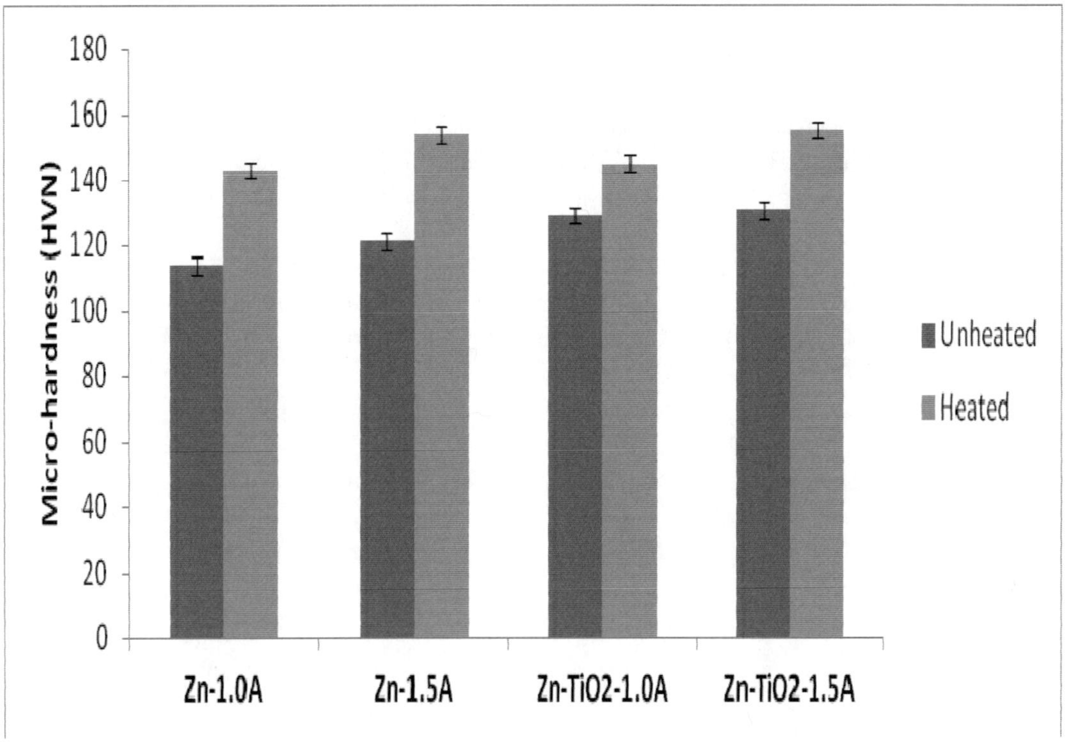

Figure 7: Hardness properties of varied current of Zn-TiO_2 nano-composite deposited on low carbon steel after heat treatment.

5. Conclusion

1. Nano structured TiO_2 particulates were used to produce Zn-TiO_2 nanocomposite coating from chloride bath.
2. The incorporation of TiO_2 in the coating was confirmed by EDX
3. The incorporation of the TiO_2 Nanoceramics particles in the zinc matrix as reinforcement enhances the hardness properties of the substrate.
4. The hardness properties increases with increase in deposition current

Acknowledgement

The authors acknowledge the Equipment support from Surface Engineering Research Centre, (SERC) Tshwane University of Technology, Pretoria, South Africa.

Reference

[1] Durodola, B.M. Olugbuyiro, J.A.O. Moshood, S.A. Fayomi, O.S.I. and Popoola. A.P.I. *Int. J. of Electrochem. Sci.*, 6, (2011) 5605

[2] Amuda, M.O.H. Subair, W.and Obitayo. O.W. *Int. J. of Eng. Research in Africa*, 2(2009)31.

[3] Basavanna, S. and Arthoba Naik, Y. *J. of Appl. Electrochem.*, 39 (2009) 1975

[4] Inegbenebor,. O.A. *J. of Mat. and Env. Sci.*, 3, (2011) 271.

[5] Popoola, A.P.I. Fayomi O.S. and Popoola., O.M. 2011. *Proc. of Mat. Sci. & Tech. Conf., Ohio:*

USA. (2011) 393-400.

[6] Rahman, M.J. Sen, S.R. Moniruzzaman, M. and Shorowordi, K.M. *J. of Mech. Eng, Tran.*, 40, (2009) 9.

[7] Deuis, R.L. Subramanian, C. and Yellup, J.M. 1996. Wear, 201, 132-144.

[8] Chen, Y. and Wang. H.M. *Surface and Coatings Technology*, 168, (2003) 30-36.

[9] Popoola, A.P.I. Pityana, S.L and Popoola. O.M. *J. of the South Afri. Inst. of Min and Met.,* 111 (2011) 345.

[10] Popoola, A.P.I. Pityana, S.L. and Popoola, O.M. *Int. J. of Electrochem. Sci.*, 6, (2011) 5038

[11] Paunovic, M. and Mordechay, S. 2nd Ed, *John Wiley and Son Inc.* 6 (2006) 388

[12] Paunovic, M. Clevenger, L.A. Gupta, Jr. Cabral, C. and Harper. J.M.E. *J. of Electrochem. Sci.* 140 (1993) 2690.

[13] Volinsky, A.A. Vella, J. Adhihetty, I.S. Sarihan, V.L. Mercado, L. Yeung, B.H. and Gerberich, W.W. *Mat. Res. Soc.*, 649, (2001)1.

[14] Popoola A.P.I and Fayomi O.S. *Int. J. of Electrochem. Sci.*, 6, (2011)3254

[15] Praveen, B.M and Venkatesha, T.V. *Int J. of Electrochem*, 261, (2011) 407

[16] Shivakumara, S. Manohar, U. Arthoba Naik, Y.& Venkatesha, T. U.. *Bull. Mat. Sci.* 30 (2007) 455

[17] Sundararajan, G. and Rama Krishna. L. *Surf. and Coat. Tech*, 167, (2003) 269.

Quaternary Zn-Ni-MoO_2-NbO_2 intermetallic composite deposits on mild steel via electrodeposition route

T. Monyai[a]*, O.S.I. Fayomi[a,b], A.P.I. Popoola[a]

[a]*Department of Chemical, Metallurgical & Materials Engineering, Tshwane University of Technology, P.M.B X680, Pretoria 0001, South Africa*
[b]*Department of Mechanical Engineering, Covenant University, P.M.B. 1023, Canaan land, Ota, Nigeria*
* Corresponding author. Tel.: +2712-382-4663; +2761-227-3095;
E-mail address: tshili.monyai@gmail.com, ojosundayfayomi3@gmail.com

Abstract- In majority of industrial applications zinc electrodeposition has been widely used for the corrosion resistant coating on steel due to its unique properties, low cost and sacrificial nature. Steel remains the most widely used material in most applications due to its excellent properties but its faults cannot be over emphasis in term of deterioration. A thin film of Zn on steel substrates was prepared by electrodeposition technique using incorporated rare earth metallic composites to form an optimised bath plating solution. Quaternary Zn-Ni-MoO_2-NbO_2 alloy coatings, as a potential replacement for Zn-Ni coatings were deposited from an acid sulphate solution. The mechanical wear and corrosion behaviours of Zn-Ni-MoO_2-NbO_2 Coated Mild Steel in 3.65% NaCl are described; these were studied by means of MTR-300 dry abrasive wear and polarization measurements respectively. The crystal particles present were observed by X-ray diffraction pattern (XRD) and energy dispersive X-ray diffraction spectrometer (EDS). The microhardness of deposited plate were investigated by means of Vickers microhardness. Scanning electron microscope and was used to study the surface morphology, and synergistic influence content in bath on surface morphology revealed that morphological changes observed in presence of additives exhibits synergistic influence on improving corrosion resistance and hardness of the deposit.
***Keywords*:** Inter-metallic, Tafel plot, Micrographs, Zn-Ni-MoO_2-NbO_2

1. Introduction

Electroplating technology is one of the long-standing surface treatment techniques employed in most industries. This method is useful due to some distinctive advantages such as affordability, operation plainness and coating quality [3-6]. Zn has been functionally coated by different methods and most research studies have focused on Zn electrodeposition [1-6]. This is because of the sacrificial protection that Zn offers to steel components [4]. However the inflexible [7] properties of Zn coatings necessitates further techniques in pursuit for improvement. In the past thick coatings were opted for as an economical enhancement of such but as technology was advancing Zn-M (where M is the iron group metals) alloy coatings were introduced [3]. Zn-Ni co-deposition is promising for obtaining good corrosion resistance and enhanced mechanical properties compared to other Zn-M coatings [1]. For this reason Zn-Ni coatings have been exploited in different industries such as automotive, construction, household, aerospace etc. [4,8].

Content from this work may be used under the terms of the Creative Commons Attribution 3.0 licence. Any further distribution of this work must maintain attribution to the author(s) and the title of the work, journal citation and DOI.
Published under licence by IOP Publishing Ltd

Molybdenum oxide additive improve properties of metallic coatings including hardness, wear resistance and corrosion resistance; it can be co-deposited especially with iron group metals [8-9]. The melting and boiling temperatures of zinc and molybdenum are different and in a study by [8] it was stated that electrodeposition can be one of the ways to obtain such environmentally friendly alloys. The present study aims to prepare Zn-Ni-MoO_2-NbO_2 quaternary inter-metallic coatings onto mild steel and evaluate and study the microstructure, micro-hardness, wear and corrosion resistance properties. To the best of author's knowledge, no work was reported on the quaternary Zn-Ni-MoO_2-NbO_2 intermetallic composite deposits on mild steel by electrodeposition route.

2. Experimental

Table 1: Compositions of the electrolyte

Composition	Mass Concentration (g/l)
$ZnSO_4$	150
Na_2SO_4	10
H_3BO_3	20
$(NH_4)_2SO_4$	20
Glycine	10
Nickel powder	60
pH	5
Voltage	0.5V and 1.0V
Time	20mins

Table 1 shows the compositions of the electrolytic bath of Zn-Ni alloy optimized without using any additives. Zn–Ni–MoO_2–NbO_2 alloy coatings were deposited from a sulphate bath with the addition of MoO_2 and NbO_2 in varying contents. The deposition operating parameters and additive contents under optimal conditions are shown in table 2. The plating solutions were freshly prepared from de-ionised water and analytical grade reagents. The pH of the electrolytes was adjusted to 4.8. The coatings were deposited on an exposed surface area of (4×4×0.02) cm of the disc electrodes made of mild steel (sourced in

IOP Conf. Series: Materials Science and Engineering **391** (2018) 012003 doi:10.1088/1757-899X/391/1/012003

Johannesburg, South Africa) which served as a cathode. The anode used was two pure Zn electrode (99.9%) with a surface area of (8×4×2) cm and both were placed at a maintained distance of 5 cm away from the cathode on the centre of the 250 mL deposition vessel. Before each experiment the surface of electrode was mechanically polished with successive grades of abrasive paper, degreased and rinsed with de-ionised water. Electrodeposition was done in room temperature (25°C) while stirring at 250 rpm for 20 minutes with an adjustable DC power source at Voltage of 0.5 to 1.0 V.

Table 2: electrolyte Summarized Bath Formulation ($Zn–Ni–MoO_2–NbO_2$)

Sample order	Matrix	% Weight MoO_2	Time of deposition (min)	Voltage
As-Received	-	0	-	-
1	$Zn\text{-}Ni\text{-}5NbO_2\text{-}5MoO_2$	10	20	0.5
2	$Zn\text{-}Ni\text{-}5NbO_2\text{-}5MoO_2$	10	20	1.0
3	$Zn\text{-}Ni\text{-}10NbO_2\text{-}10MoO_2$	15	20	0.5
4	$Zn\text{-}Ni\text{-}10NbO_2\text{-}10MoO_2$	15	20	1.0

Surface morphology of the coatings was characterized using scanning electron microscopy (SEM, JSM–7610F), energy dispersive spectroscopy (EDS) and Optical Microscope. The wear resistance towards abrasion was evaluated using the modification of MTR-300-ASTM-G-65 dry sand rubber wheel abrasion test on prepared sample of 10x10mm. Micro-hardness studies were carried out using a Diamond pyramid indenter EMCO Test Dura-scan micro-hardness testers at a load of 10 g for a period of 20 s before and after heat treatment at 250 °C and 350 °C. The electrochemical behaviour of intermetallic coatings was monitored in 3.65 wt.% NaCl solution by potentiodynamic polarization using Linear Polarization Resistance (LPR) techniques; the equipment used was metrohμ AUTOLAB electrochemical workstation utilizing NOVA software. All electrochemical measurements were made using three electrode set up with developed coating (1 cm^2 exposed surface area) as working electrode, graphite as counter electrode and saturated calomel electrode (SCE) as reference. The corrosion currents were obtained from the tafel's extrapolation method.

IOP Conf. Series: Materials Science and Engineering **391** (2018) 012003 doi:10.1088/1757-899X/391/1/012003

3. Results and discussion

3.1. SEM/EDS analyses of deposition

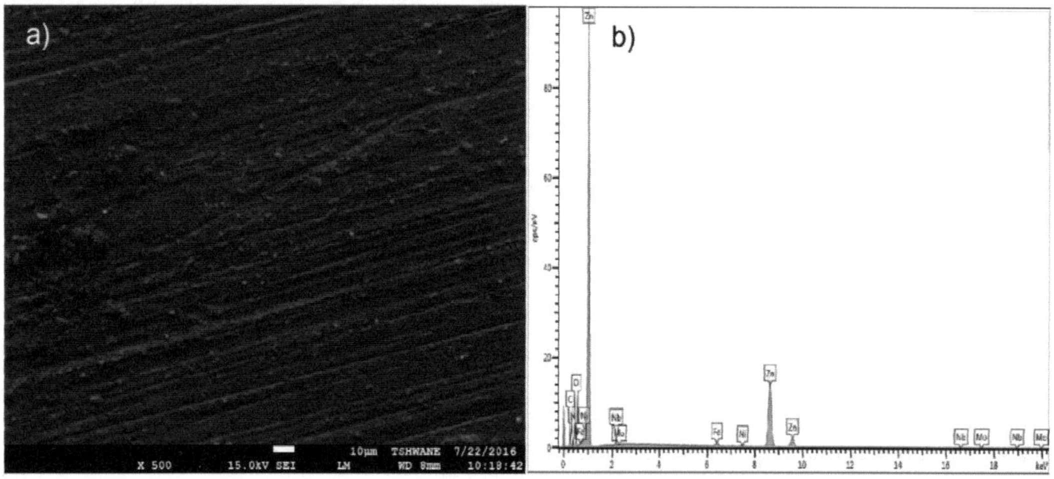

Figure 1: SEM/EDS Spectra showing the Surface Morphology of Zn-Ni-5NbO$_2$-5MoO$_2$-1.0V

Figure 1 and 2 shows the SEM-EDS spectra of the Zn-Ni-5NbO$_2$-5MoO$_2$-1.0V and Zn-Ni-10NbO$_2$-10MoO$_2$-1.0V intermetallic composite coating deposited for 20 minutes respectively. The EDS pattern of the developed coatings confirms the crystalline nature of the thin film developed while showing the elemental presence (in appreciable amounts) of the induced MoO$_2$ and NbO$_2$ with visible peaks for both the samples analysed.

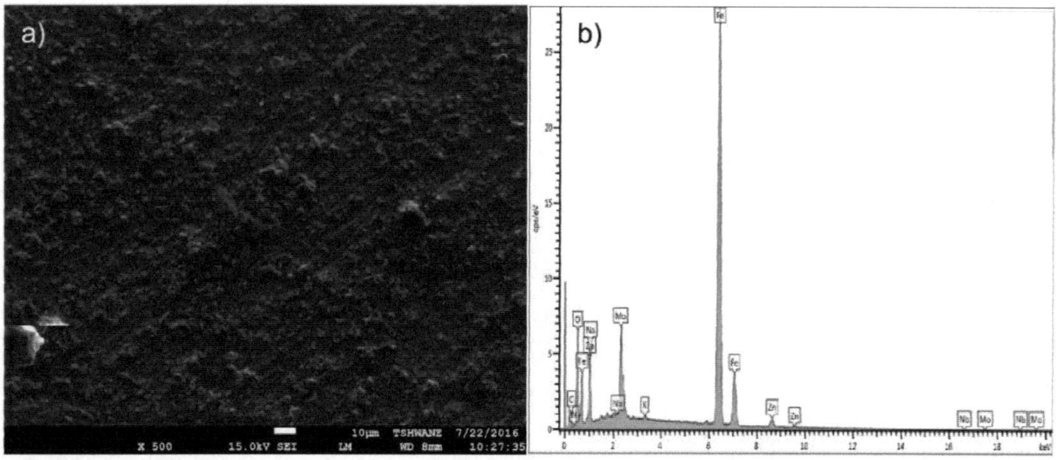

Figure 2: SEM/EDS Spectra showing the Surface Morphology of Zn-Ni-10NbO$_2$-10MoO$_2$-1.0A Sulphate Deposition

Generally, good surface appearances and smooth orientation was obtained as a result of the proposed quaternary enhancements. Uniform arrangement of the surface was noticed with sprinkles of dispersed particles. Much smaller scattering grains were noticed for increased content Zn-Ni-NbO_2-MoO_2deposit in figure 2 (a). This might be due to the increased ions [2] of MoO_2 and NbO_2 on the morphology and thus sows surface modifications. No visible pores or cracks were revealed on the interface of the coatings and therefore one can state that good adhesion was obtained for the optimized deposition parameters in the study.Therefore the presence of molybdenum and niobium oxide has a significant effect on obtained smooth morphology of the alloy coating surface.

3.2. Micro-hardness Behavior

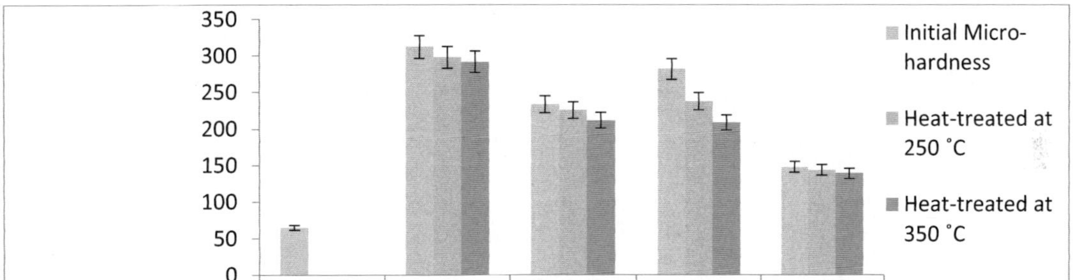

	As-received	Zn-Ni-5NbO2-5MoO2-0.5V	Zn-Ni-5NbO2-5MoO2-1.0V	Zn-Ni-10NbO2-10MoO2-0.5V	Zn-Ni-10NbO2-10MoO2-1.0V
Initial Micro-hardness	65	312	234	282	148
Heat-treated at 250 °C	0	298	226	238	144
Heat-treated at 350 °C	0	292	212	209	139

Figure 3: Micro-hardness Chart for Zn-Ni-MoO_2-NbO_2

Figure 3 shows the micro-hardness variations of the deposited coatings against the as-received sample for different temperature subjection (25, 250 and 350 °C) respectively. Alloying zinc coatings with nickel can enhance their corrosion resistance, micro-hardness and thermal stability [7], this might be even better for co-deposition with refractory oxide particles such as MoO_2 and NbO_2. The presence of MoO_2 and NbO_2 improves the micro-hardness [4] for all the co-deposited coatings obtained, but it has negative influence on the corrosion and wear for the Zn-Ni-10MoO_2-10NbO_2-0.5V sample proving to be the unstable sample by a sharp decrease (44 HV) after heat treatment. The thermal diffusion characteristics were analysed by heat treatment (at 250 °C and 350 °C) subjection and the micro-hardness variations were evaluated. Highest wear results were obtained for Zn-Ni-10NbO_2-10MoO_2-1.0V this can be explained by the lowest micro-hardness obtained, therefore it can be attested that the micro-hardness values obtained affirms both the wear and corrosion results. This statement is further confirmed by sample Zn-Ni-5NbO_2-5MoO_2-1.0V which proved to be the best sample in the deposition matrix in terms of corrosion, wear and micro-hardness. Generally as the temperature was increased there was a slight decrease in the micro-hardness values. These Zn-Ni alloyed coatings can be useful for fasteners of car engines, because they retain their properties at temperatures variations [7].

3.3. Electrochemical Behaviour of Deposition

Table 3: Potentiodynamic Polarization Data

Sample	I_{corr} (A/cm^2)	$R_P(\Omega)$	E_{corr} (V)	Corrosion rate (mm/yr)
Control	0.038119	11.708	-1.3065	442.94
Zn-Ni-5NbO$_2$-5MoO$_2$ -0.5V	0.002557	66.787	-1.3867	29.717
Zn-Ni-5NbO$_2$-5MoO$_2$ -1.0V	0.000159	471.1	-1.0286	11.8422
Zn-Ni-10NbO$_2$-10MoO$_2$ -0.5V	0.004758	31.71	-0.95087	55.288
Zn-Ni-10NbO$_2$-10MoO$_2$ -1.0V	0.003523	52.692	-0.60053	40.932

Data on the corrosion potential (Ecorr), corrosion current density (icorr) and corrosion rate (CR) and polarization resistance (Rp) of Zn–Ni-NbO$_2$-MoO$_2$ alloys is summarized in Table 3.

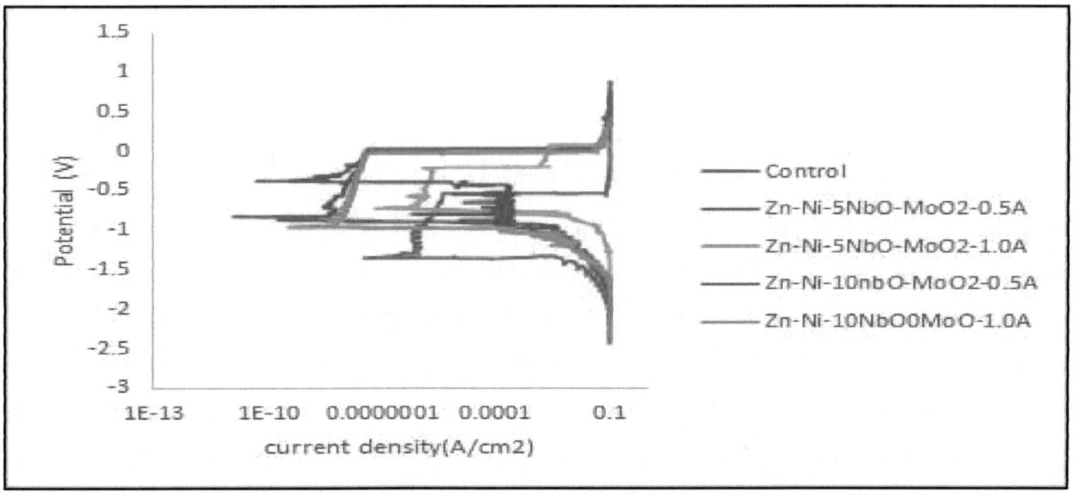

Figure 4: Potentiodynamic Polarization Curves for Zn-Ni-NbO$_2$-MoO$_2$ Deposition

The polarization curve of the Zn-Ni-NbO_2-MoO_2 matrix deposition is shown in figure 4. The enhanced corrosion behaviour of the quaternary alloy deposited is attributed to the chemical composition and the synergistic effect of MoO_2 and NbO_2. Again additive increase seemingly increases with the corrosion rate thus reducing the polarization rate. In the case of deposited coatings the lowest obtained corrosion rate was 11.8422 mm/year for Zn-Ni-5NbO_2-5MoO_2 characterized coatings deposited at 1.0V and this implies that the deposition parameters resulted in strongest passive layer on the coating's surface. It is known from literature [2] that MoO_2 co-deposition enhances the properties of Zn or rather iron group metallic coatings. Generally it can be said that there is a passive film on all the coatings but there is a possibility of passitivity breakdown for Zn-Ni-10MoO_2-10NbO_2 and Zn-Ni-5NbO_2-5MoO_2 both deposited at 0.5 V.

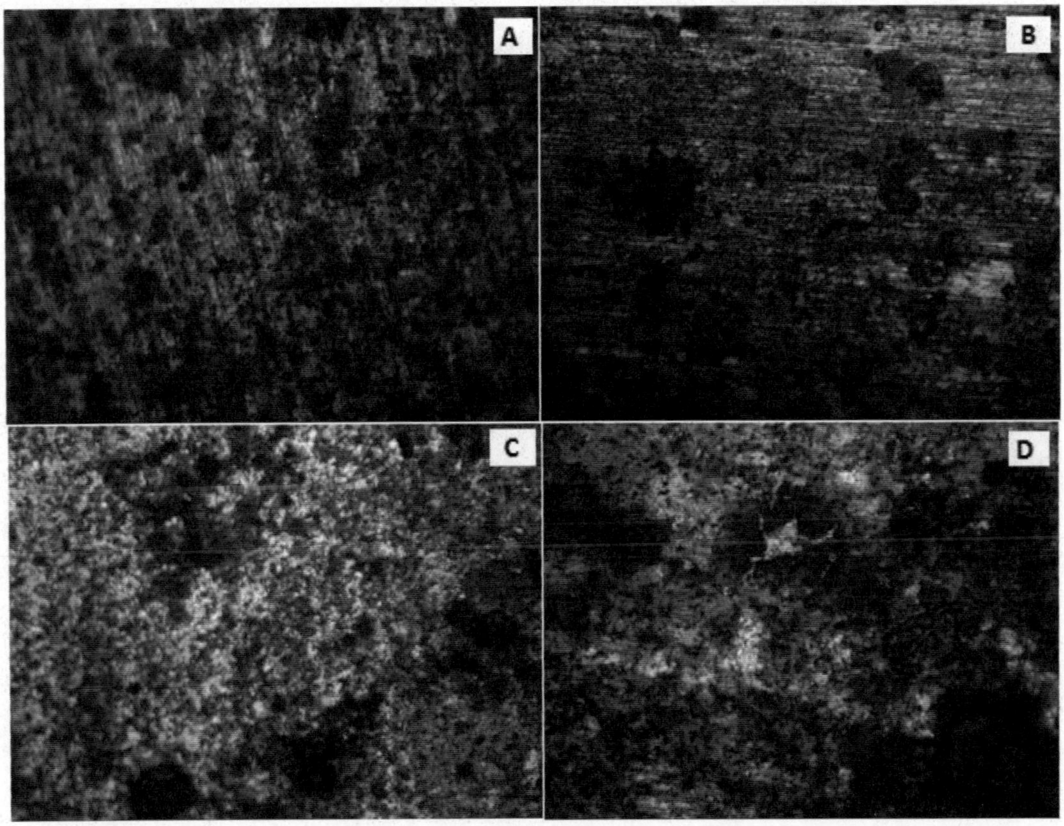

Figure 5: Optical Micrograph of (a) Zn-Ni-5NbO_2-5MoO_2 at 0.5A (b) Zn-Ni-5NbO_2-5MoO_2 at 1.0A (c) Zn-Ni-10NbO_2-10MoO_2 at 0.5A (d) Zn-Ni-10NbO_2-10MoO_2 at 1.0V after corrosion

From all indications, uniform attack by aggressive chloride anions can be seen from figure 5 which shows the optical micrographs obtained after corrosion for all the coating. The corrosion resistance of these coatings depends on the additives [7] contents and deposition parameters. The corrosion reactions rendered the rough surface observed.

IOP Conf. Series: Materials Science and Engineering **391** (2018) 012003 doi:10.1088/1757-899X/391/1/012003

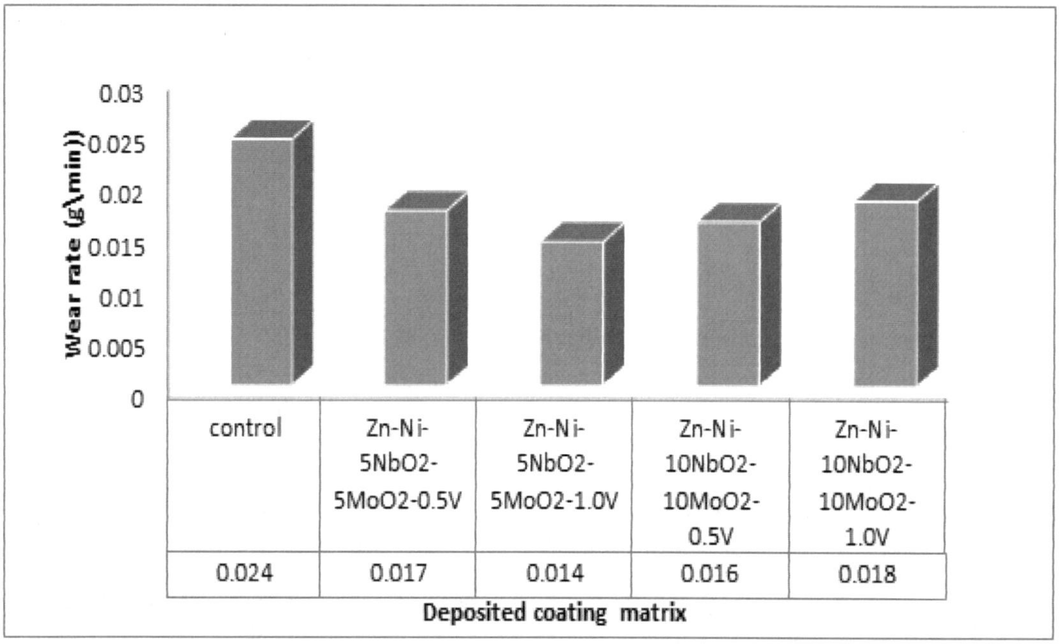

Figure 6: Variation of the wear rate with time (Zn-Ni-NbO_2-MoO_2deposition)

The wear loss ranges from the highest to the lowest is indicated by: Zn-Ni-10NbO_2-10MoO_2-1.0V Zn-Ni-5NbO_2-5MoO_2-0.5V, Zn-Ni-10NbO_2-10MoO_2-0.5 V $_{Zn}$-Ni-5NbO_2-5MoO_2-1.0V. This is graphically represented in figure 6. Higher voltage of 1.0V yielded both the highest and lowest wear loss with time. Correct adjustment could distinguish the better performing sample this can be attributed to a statement by [3] about specific bath additives.

4. Conclusions

In conclusion the Zn-Ni-NbO_2-MoO_2 intermetallic matrix was successful deposited. The induced co-deposited oxides synergistically improved the corrosion, wear, micro-hardness and surface properties of mild steel. Energy Dispersive Spectroscopy results confirmed the deposition of both NbO2 and MoO2 on the surface interface. The coated samples prove to retain thermal stability at temperature variations.

Acknowledgement

Appreciation goes to the National Research Foundation for financial assistance and Surface Engineering Research Centre (SERC) Tshwane University of Technology, Pretoria for equipment support.

References

[1] B. Szczygieł, A.Laszczyńska, and W. Tylus. 2010), Surf & Coat.Tech, **204** 1444 (2010).

[2] N. Eliaz, K. Venkatakrishna, and A. Chitharanjan Hegde, Surf. & Coat. Tech, **205**, 1978(2010),
[3] S. Rashmi, L. Elias, A. Chitharanjan Hegde. Eng Sci. and Tech, (2016)
[4] G.H.Barati Darband, A. Afshar, and A. Aliabadi, Zn–Ni Surf & Coat Tech, (2015).
[5] O.S.I. Fayomi, A.P.I. Popoola and T. Monyai Bull. Chem. Soc. Ethiop., **30**(3), 449-456 (2016)
[6] V.G. Roev, R.A.Kaidrikov and A.B. Khakimullin, Russ. Jour. of Elec. **37** (7), 756 -759(2001)
[7] S. Ranganatha, T.V. Venkatesha & K. Vathsala, Appl. Surf. Sci. **256** 7377–7383 (2010)
[8] H. Kazimierczak, P. Ozga, Z. Swiatek and E. Bielanska, Jour.of Alloys and Comp. **578** 82–89 (2013)
[9] E. Gomez, E. Pellicer and E. Valles, Electro. Comm. **7** 275–281(2005)
[10] M.J. Zhao, C. Cai, L. Wang, Z. Zhang, and J.Q. Zhang, Surf & Coat.Tech. **205** 2160-2166 (2010)

IOP Conf. Series: Materials Science and Engineering **391** (2018) 012004 doi:10.1088/1757-899X/391/1/012004

Oil as Corrosion Inhibitor for Aluminium Alloy in Aggressive Environment

O .SANNI, API POPOOLA, OSI FAYOMI

Department of Chemical, Metallurgical and Materials Engineering, Tshwane University of Technology, Pretoria, South Africa
Corresponding author: tayo.sanni@yahoo.com

ABSTRACT. Corrosion processes are responsible for huge losses in industry. This deterioration of components results in loss of plant efficiency, total shutdown and aggressive damage in a number of industries. Though organic, inorganic and mixed material inhibitors were used for a long time to combat corrosion, the environmental toxicity of inorganic corrosion inhibitors has prompted the search for organic corrosion inhibitors. In search for this, the use of organic inhibitors is one of the best options of protecting metals and alloys against corrosion as it is environmentally acceptable, inexpensive and readily available. In this regards, silicon oil was used as inhibitor in this study. Weight loss, corrosion rate, inhibition efficiency and potentiodynamic polarization techniques were used in this study. The aluminium alloy samples exposed to the inhibitor showed a lower corrosion rate values and excellent polarization resistance as compared with the corrosion rate samples without inhibitor. The corrosion inhibition rate increases with increasing the concentration of the silicone oil. The adsorption isotherm was confirmed by Langmuir adsorption isotherm.
Keywords: Corrosion; Silicone oil; Inhibition; and Corrosion rate

1. INTRODUCTION

Aluminum comprises eight percent of the earth's crust. The annual market of aluminium is 25 million tons which makes it one of the principal nonferrous metals in the metallurgy industry [1]. The applications of aluminium and its alloys varied from household to electrical engineering, packaging, transport and building [2-6]. Also, it demonstrates a good fatigue resistance as well as high strength to weight ratio. Aluminium resistance to corrosion in aqueous media can be attributed to a rapidly formed surface oxide film. Because the aluminium oxide layer is very thin, it is easily degraded by physical or chemical methods. On the contrary, this alloy is highly susceptible to corrosion because of the presence of inter metallic particles at the surface resulting in need for the development of several methods in order to enhance its resistance against corrosion [7, 8]. Efforts have been made to protect the corrosion of aluminium surface in different corrosive environments. Recently, the addition of inhibitors has been considered to be one of the most common approaches to hinder the corrosion of aluminium [9-13]. Aliphatic, aromatic amines and nitrogen heterocyclic molecules compounds have been widely reported as corrosion inhibitors of aluminum in aggressive media [14-16]. However, some of these compounds are costly and not environmentally friendly. In search for an alternative, environmentally friendly, protection for alluminium alloy, the present study aims to investigate the inhibition action of silicone oil and their protective

Content from this work may be used under the terms of the Creative Commons Attribution 3.0 licence. Any further distribution of this work must maintain attribution to the author(s) and the title of the work, journal citation and DOI.
Published under licence by IOP Publishing Ltd

IOP Conf. Series: Materials Science and Engineering **391** (2018) 012004 doi:10.1088/1757-899X/391/1/012004

performance for aluminium alloy in 3.5% NaCl solution using a variety of electrochemical tests; weight loss, corrosion rate, inhibition efficiency and potentiodynamic polarization techniques. Adsorption mechanism was also investigated.

2. MATERIALS AND METHODS

2.1. Materials and Sample Preparation

The aluminum alloy sheet with chemical composition shown in Table 1 below was mechanically press-cut into specimens of dimension 30 mm by 30mm by 0.3mm. All test plates of aluminium alloy samples were ultrasonically degreased in acetone for 15 min. After that, the aluminum alloy specimens were washed thoroughly with distilled water and dried. The specimens with an exposed area of 1 cm^2 were used for potentiodynamic polarization tests. Analytical reagent grade of sodium chloride and double distilled water were used to prepare the corrosive medium. Molecular structures of silicone oil are shown in Figure1.

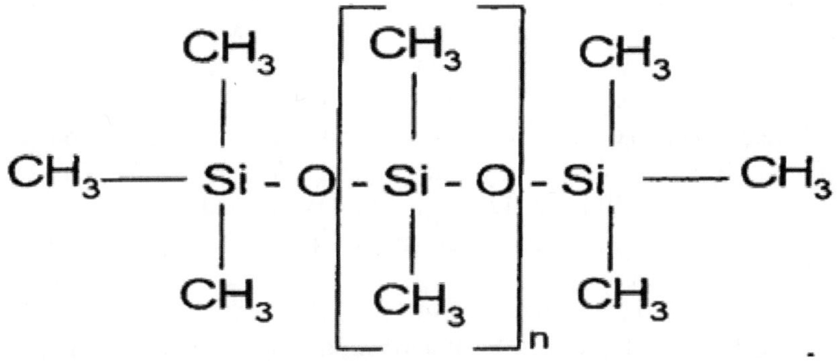

Figure 1. Molecular structure of silicone oil

2.2. Weight Loss Test

The aluminum alloy specimens in triplicate were immersed in a 3.5 % NaCl solution with and without different concentration (0.1–0.5 g/L) of silicone oil for 168 hours. The temperature was controlled by a thermostat aqueous bath at 304K. For all weight loss tests, the volume of prepared solution was 250 mL. After the immersion, all specimens were brought out from the solution, scrubbed with bristle brush under running water, then washed thoroughly with distilled water, dried in acetone, and weighed accurately. Minimum of three samples were tested and the average value was obtained. The corrosion rates and inhibition efficiency were determined from equations previously reported by [13].

Table 1: Chemical composition of aluminium alloy (%wt)

Element	Compositions
Si	0.157
Fe	0.282
Cu	0.0025

Mn	0.024
Mg	0.51
Cr	0.023
Ti	0.006
Ca	0.0011
Zr	0.002
V	0.0035
Al	Balance

2.3. Electrochemical Measurements

The potentiodynamic polarization tests were carried out in a three-electrode cell system with a non-aqueous Ag/Ag electrode used as reference electrode, and a platinum wire used as counter electrode. The working electrode was mounted aluminium alloy sample. For each sample, measurements were performed in at least triplicate. The following electrochemical parameters were determined: corrosion current density (icorr), corrosion potential (Ecorr), and corrosion rate (CR). The corrosion inhibition effectiveness, IE, was calculated accordingly using equation reported elsewhere. All specimens were embedded in Teflon using epoxy resin and were prepared as rotating disc electrodes. Before the electrochemical measurements, all samples were mechanically polished with 1,200 grit emery paper, washed with double distilled water and put into the cell containing the studied solutions. There was an immersion in the solution for 1 h for the open circuit potential to reach a steady state. The polarization curves were obtained by changing potential from -250 mV to +250 mV with a scan rate of 0.5mV/s. The electrochemical experiments were performed using a Potentiostat/Galvanostat with NOVA software, Version 1.8. Each experiment was repeated at least three times to ensure the reproducibility.

2.4 Adsorption Isotherm

To understand the adsorption mechanism of the corrosion resistance of aluminium alloy surface, the data obtained from the corrosion study were tested with several adsorption isotherms models including Langmuir, Frumkin, and Temkin. The surface coverage, θ, was calculated using equation 1 below where η is the inhibition efficiency:

$$\theta = \frac{n(\%)}{100(\%)} \qquad (1)$$

3. RESULTS AND DISCUSSION

With the silicone oil concentration varying from 0.1 to 0.5 g/L in saline solution, the weight loss by the aluminium alloy in various concentration of silicone oil took the order of Control < 0.1 < 0.2 < 0.3 < 0.4 < 0.5.

3.1. Adsorption Characteristics

Inhibition efficiency of silicone oil falls within the first 96 hour but becomes linear after 120 hours. For the period of 96 hours the inhibition efficiency of silicone oil was averagely maintained at 70% and 83% respectively at 0.4 and 0.5g/L of silicone oil as indicated in Figures 2, 3 and 4. Corrosion rate of aluminum in various concentrations of silicone oil in the studied medium are in the order control > 0.1 > 0.2 > 0.3 > 0.4 > 0.5 as shown in Figure 4. As corrosion of aluminum alloy increases with time that of inhibitor decreases with inhibitor concentration. It was observed that out of the five concentrations of silicone oil that were tested for weight loss, 0.5g/L gave the highest efficiency. This implies that 0.5g/L concentration of silicone oil inhibited the aluminum alloy most and followed by 0.4g/L. The silicone oil contents are expected to have affinity to bind with the aluminum alloy surface. Therefore elements contained in this oil are required to be attracted to aluminum alloy coupon and adhered to its surface by chemical reactions thereby preventing direct attack by the medium.

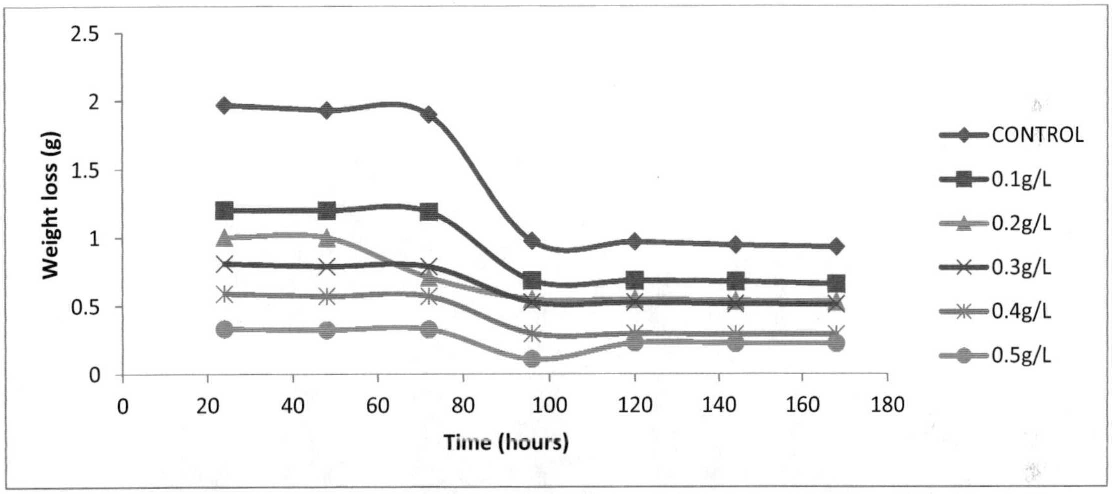

Figure 2: Weight loss of aluminium alloy with time of immersion in 3.5% NaCl at 304K at different inhibitor concentration

Figure 4 indicated the higher inhibition efficiency potential of silicone oil to be higher with increase in its concentration. In Figure 3, corrosion rate of the samples with inhibitor decreased above the control at all the concentrations studied. The adsorption of this oil on the aluminium alloy surface reduces the rate of corrosion. It shows aluminium alloy has affinity for elements that made up its ore rather than organic constituents of silicone oil. The addition of the oil hindered the affinity of the aluminium alloy for the elemental components of the oil thereby reduced the adsorption of the components on the metal. Different oil has been reported to inhibit corrosion of metals [18-23]. Due to interaction between the metal and the oil there could be formation of a soluble compound which may stimulate corrosion rather than favor inhibition. Also the properties responsible for effective inhibition such as the number of adsorption active centers in the molecule, their charge density, the molecular size and the mode of adsorption may have been restructured. The ability of the silicone oil to reduce aluminium corrosion is shown in Figures 2, 3 and 4

This confirms that pure metals are ready to return to their ore or oxide state in the presence of required substances and in the absence of organic substance from oil with significant sodium content will retard adsorption of inhibitors on the metal surface.

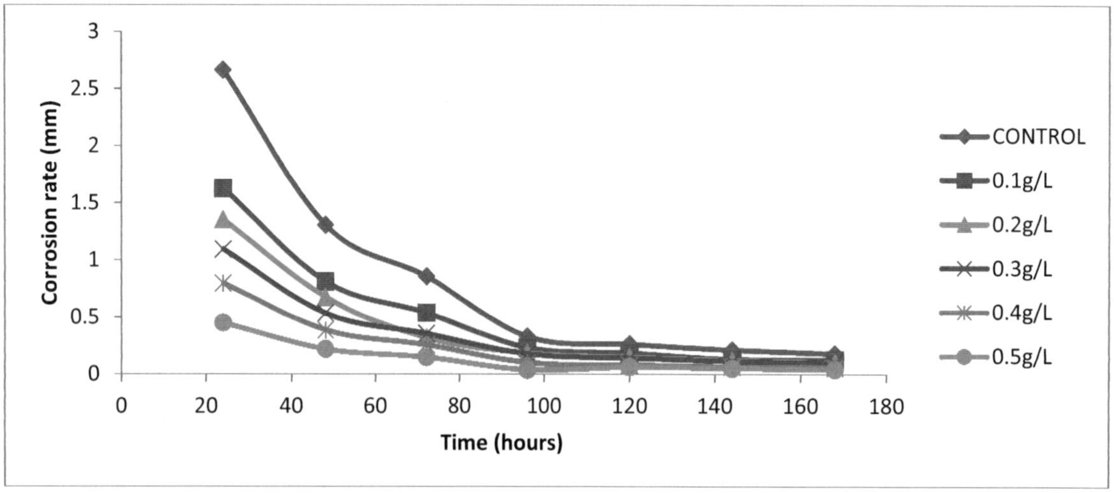

Figure 3: Corrosion rates of aluminium in 3.5% NaCl at 304K at different inhibitor concentration

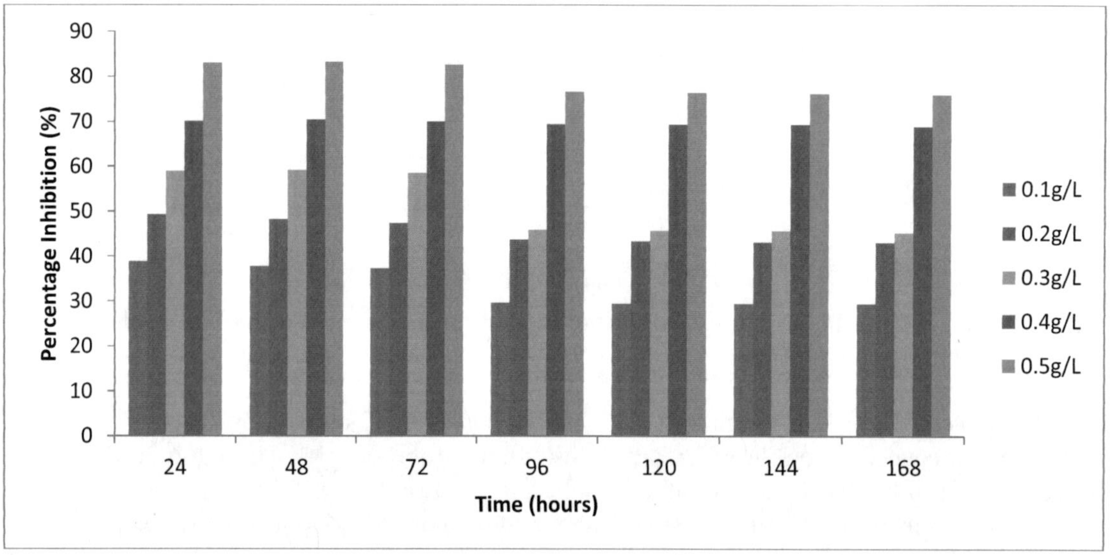

Figure 4: Percentage inhibition efficiency of aluminium in 3.5% NaCl at 304K at different inhibitor concentration

Table 2. Data obtained from Tafel plot curves of aluminium alloy in 3.5%NaCl solution at different concentrations of silicone oil

S/N	C (g/L)	Icorr (A/cm2)	βc (V/dec)	βa (V/dec)	LPR Rp (Ωcm2)	-Ecorr (V)	CR (mm/yr)
1	Control	0.000802	0.60085	1.3086	175.96	-0.92732	3.7570
2	0.1	0.000235	0.27822	0.64902	359.87	-0.92732	2.7309
3	0.2	0.000164	0.02582	0.1081	86.79	-0.92732	1.9011
4	0.3	0.000156	0.20144	0.094419	178.57	-0.92732	1.8168
5	0.4	5.75E-05	0.24695	0.352	196.4	-0.92732	0.6679
6	0.5	4.28E-05	0.16668	0.15949	55.324	-0.92732	0.4974

Current density-potential curves were recorded to obtain information on the effectiveness of the silicone oil on the corrosion processes and its effect on the anodic and cathodic reactions. Figure 5 represent the anodic and cathodic polarization curves for the aluminium alloy immersed in 3.5% NaCl in the absence and presence of different concentrations of silicone oil at 304K. From these polarization curves the corrosion kinetic parameters such as corrosion current density (icorr), corrosion potential (Ecorr), anodic and cathodic Tafel slopes (ba, bc) were deduced by extrapolating Tafel's lines.

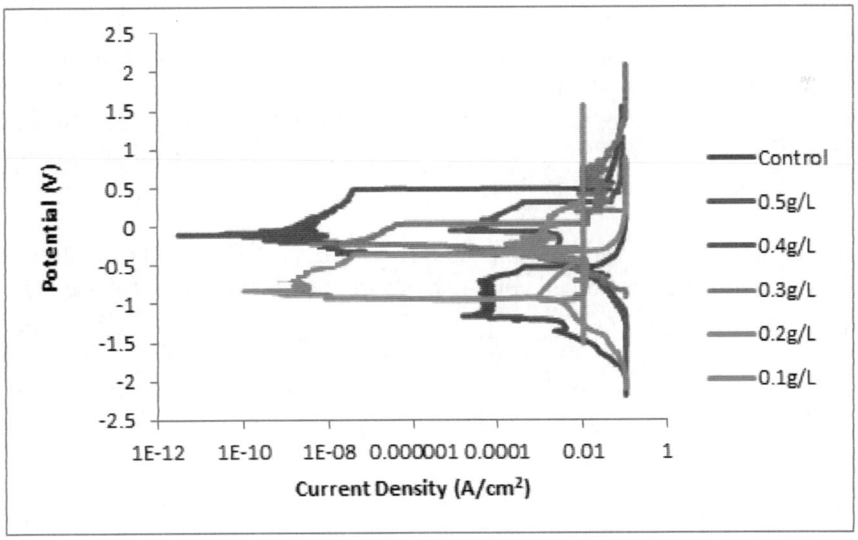

Figure 5. Tafel polarization curves of aluminium alloy in 3.5% NaCl solution in the absence and presence of different concentrations of silicone oil

The changes observed in the polarization curves in the presence of the inhibitor are usually used as criteria to classify inhibitors as anodic, cathodic or mixed. Figure 4 shows the shifting of corrosion potential (Ecorr) as dependent on inhibitor concentrations, while the general shape of anodic and cathodic curves in the absence and presence of the inhibitor is comparable. The curves are shifted towards lower current density values compared to that of the control solution, indicating the inhibition of the aluminium alloy. The decrease in corrosion rate associated with a shift of both cathodic and anodic plots of the polarization curves toward lower current densities, and a positive shift in corrosion potential, indicates that this oil acts as a mixed-type inhibitor. When the concentration of oil increases, the anodic branches make a sudden shift to more positive potential direction. The data in Table 2 show that the bc values are almost constant at different concentrations of inhibitor. This observation indicates that the mechanism of proton discharge reaction is not affected in the presence of silicone oil. Likewise, the constancy of the cathodic slope may have indicated that the hydrogen evolution reaction is activation controlled and that the inhibitor is absorbed by simply blocking the active sites of the aluminium surface [17]. In the anodic area, as the inhibitor concentration increases, there is a shift towards positive values of corrosion potential. However, the slight variation of ba near the shift of Ecorr toward anodic potentials may be attributed to the modification of the anodic process due to the formation of a protective film on the electrode surface rather than a simple adsorption on the active sites. The analysis of these results shows that the increase of silicone oil concentration decreases icorr, and consequently there is an increase of inhibition efficiency. The increase in efficiency may be due to the blocking effect on the surface by both adsorption and film formation mechanism, which decreases the effective area of attack.

Adsorption Isotherm

Adsorption characteristics are helpful in understanding the nature of corrosion inhibition and it can be deduced in the term of adsorption isotherm. By fitting the various adsorption isotherms into Freundlich, Temkin, Langmuir and Frumkin isotherms, the data were tested graphically. The process of adsorption occur when liquid or gas accumulates on the surface of a solid or liquid, forming a molecular or atomic film. In general, adsorption isotherm can be divided into two categories. First category occurs when the molecules adsorb but their chemical originality remains unchanged. In this state, the adsorption bond between inhibitor molecule and surface is fairly weak. The best fitted isotherm that describes the adsorption behavior of silicone oil on aluminium surface was Langmuir adsorption isotherm and it can be expressed by using the equation as described by [23].

It was found that the Langmuir adsorption model was the best fit with correlation coefficients above 0.858 for the temperature studied (Figure 6) while the others correlated are less than 0.602.

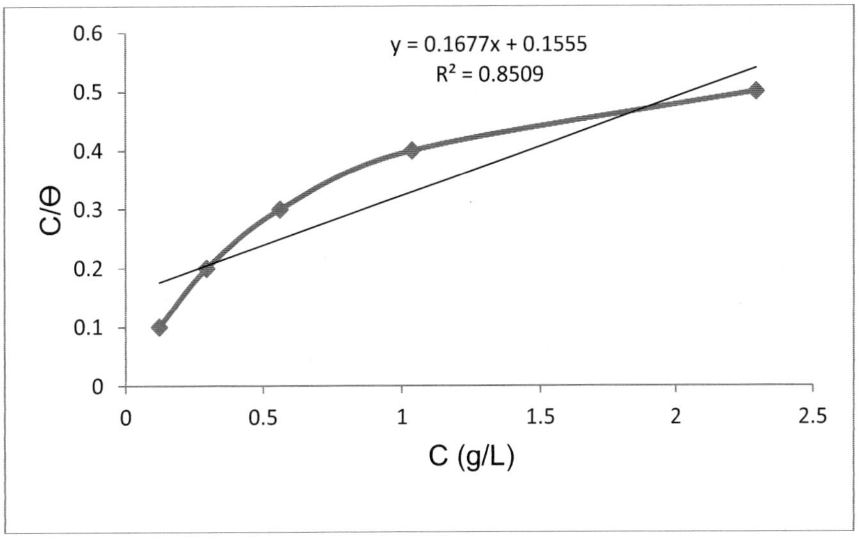

Figure 6. Langmuir adsorption isotherm model for silicone oil in 3.5% NaCl medium

4. CONCLUSION

Our study indicates that silicone oil acts as an effective corrosion inhibitor for aluminium alloy in 3.5% NaCl solution. The inhibition efficiency of the inhibitor concentration is in the order 0.5 > 0.4 > 0.3 > 0.2 > 0.1 > Control. Polarization tests indicated that this oil behaves as a mixed-inhibitor type with a predominant anodic effectiveness at higher concentration. The results obtained from weight loss measurements and electrochemical techniques were in reasonable agreement.

REFERENCES

[1] Schutze, M, Wieser D. and Bender, R. Corrosion Resistance of Aluminium and Aluminium Alloys, Wiley –VCH (2010).

[2] Gudic, S, Vrsalovic, L, Kliskic, M, Jerkovic, I, Radonic, A, and Zekic, M. 2016. Corrosion inhibition of AA5052 aluminium alloy in 0.5 M NaCl solution by different types of honey. Int. J. Electrochem. Sci., 11, 998–1011.

[3] Rosliza, R, Nik, W.B.W, Izman, S and Prawoto, Y. 2010. Anti-corrosive properties of natural honey on Al–Mg–Si alloy in seawater. Curr. Appl. Phys. 10, 923–929.

[4] Hill, J.A, Markley, T, Forsyth, M, Howlett, P.C and Hinton, B.R.W. 2011. Corrosion inhibition of 7000 series aluminium alloys with cerium diphenyl phosphate. J. Alloy. Compd., 509, 1683–1690.

[5] Rosliza, R, Nik, W.B.W and Senin, H.B. 2008. The effect of inhibitor on the corrosion of aluminum alloys in acidic solutions. Mater. Chem. Phys., 107, 281–288.

[6] Dabala, M, Ramous, E and Magrini, M. 2004. Corrosion resistance of cerium-based chemical conversion coatings on AA5083 aluminium alloy. Mater. Corros. 55, 381–386.

[7] Oubaha, M, Varma, P.C.R., Duffy, B, Gasem, Z.M, and Hinder, S.J. 2014. Development of a novel hybrid aluminum-based sol-gel materials: application to the protection of

AA2024 T3 alloys in alkaline environment, Advances in Materials Physics and Chemistry 4, 75–84.

[8] Davis, J.R. 1999. Corrosion of aluminium and aluminium alloys, ASM International, Materials Park, OH.

[9] Oguzie, E.E, Onuoha, G.N and Ejike, E.N. 2007. Effect of gongronema latifolium extract on aluminium corrosion in acidic and alkaline media. Pigment Resin Technol., 36, 44–49.

[10] Al-Turkustani, A.M, Arab, S.T and Al-Dahiri, R.H. 2010. Aloe plant extract as environmentally friendly inhibitor on the corrosion of aluminum in hydrochloric acid in absence and presence of iodide ions. Mod. Appl. Sci., 4, 105–124.

[11] Li, J, Hurley, B and Buchheit, R. 2015. Microelectrochemical characterization of the effect of rare earth inhibitors on the localized corrosion of AA2024-T3. J. Electrochem. Soc., 162, C563–C571.

[12] Ilevbare, G.O and Scully, J.R. 2001. Mass-transport-limited oxygen reduction reaction on AA2024-T3 and selected intermetallic compounds in chromate-containing solutions. Corrosion, 57, 134–152.

[13] Qian Z, Tiantian T, Peilin D, Zhiyi Z and Fang W. 2017. The Corrosion Inhibition Effect of Triazinedithiol Inhibitors for Aluminum Alloy in a 1M HCl Solution, Metals, 7, 44, 1-11.

[14] Hachelef, H, Benmoussat, A, Khelifa, A, Athmani, D and Bouchareb, D. 2016. Study of corrosion inhibition by Electrochemical Impedance Spectroscopy method of 5083 aluminum alloy in 1 M HCl solution containing propolis extract. J. Mater. Environ. Sci. 7, 1751–1758.

[15] Khaled, K.F. 2010. Electrochemical investigation and modeling of corrosion inhibition of aluminum in molar nitric acid using some sulphur-containing amines. Corros. Sci., 52, 2905–2916.

[16] Zhang, Q.B and Hua, Y.X. 2010. Corrosion inhibition of aluminum in hydrochloric acid solution by alkylimidazolium ionic liquids. Mater. Chem. Phys., 119, 57–64.

[17] El-Naggar, M. M. 2007. Corrosion inhibition of mild steel in acidic medium by some sulfa drugs compounds, Corros. Sci., 49, 2226–2236.

[18] Afia, L, Benali, O, Salghi, R, Ebenso, E.E, Jodeh, S, Zougagh, M and Hammouti, B. 2014. Steel Corrosion Inhibition by Acid Garlic Essential Oil as a Green Corrosion Inhibitor and Sorption Behavior, Int. J. Electrochem. Sci., 9, 8392 – 8406.

[19] Benabdellah, M., Hammouti, B., Bendahhou, M. and Aouniti, A. 2006, Inhibition of steel corrosion in 2M H3PO4 by artemisia oil, Applied Surface Science, 252, 6212-7.

[20] Booth, G., Delatte, D.E. and Thames, S.F. 2007, Incorporation of drying oils into emulsion polymers for use in low-VOC architectural coatings, Industrial Crops and Products, 25, 257-65.

[21] Bouyanzer, A. and Hammouti, B. 2004a, A study of anticorrosive effects of artemisia oil on steel, Pigment & Resin Technology, 33, 287-92.

[22] Bouyanzer, A. and Hammouti, B. 2004b, Naturally occuring ginger as corrosion inhibitor for steel in molar hydrocloric acid at 353 K, Bulletin of Electrochemistry, 20, 63-5.

[23] Bouyanzer, B., Hammouti, B. and Majadi, L. 2006, Pennyroyal oil from Mentha pulegium as corrosion inhibitor for steel in 1 M HCl, Materials Letters, 60, pp. 2840-3.

Effect of Powder Flow Rate on Surface Finish in Laser Additive Manufacturing Process

R. M. Mahamood[1,2,*] and E. T. Akinlabi[1]

[1]Department of Mechanical Engineering Science, University of Johannesburg, Auckland Park Kingsway Campus, Johannesburg, 2006, South Africa.
[2]Department of Mechanical Engineering, University of Ilorin, Nigeria.

Abstract. Laser metal deposition process is an additive manufacturing process that can produce three-dimensional part through material addition. It that belongs to the class of directed energy deposition. This technology is still fairly new and some of the physics of the process is not fully understood. In this study, the influence of the processing parameter, namely: the powder flow rate, on the resulting physical property of laser metal deposited titanium alloy powder on titanium alloy substrate was carefully investigated. The powder flow rate was varied between 1.44 g/min and 7.2 g/min while the laser power, the gas flow rate, the laser spot size and the scanning speed were maintained at constant values of 3 kW, 2 l/min, 2mm and 0.004 m/s respectively. The laser metal deposition experiment was conducted using a 4.0 kW Nd: YAG laser. The surface roughness was measured using Jenoptik surface analyzer. The results showed that the surface roughness increases when the powder flow rate was increased. The minimum average surface roughness value of 4.5 μm was achieved at the lowest powder flow rate of 1.44 g/min. The low roughness value obtained at lower powder flow rate could be attributed to proper melting of the powder at these settings.

Key Words: Additive manufacturing, laser metal deposition process, Nd:YAG laser, surface roughness,Ti6Al4V

1. Introduction

Laser material deposition (LMD) is an important additive manufacturing technology for producing solid object from the three dimensional (3D) computer aided design (CAD) data of the object. This is achieved by building up materials layer after layer by following the path generated by the CAD data [1]. In adaptation to the ability of the process to fabricate new 3D components, it can also be used to repair a high valued component which used to be difficult to repair [2, 3]. The LMD process also offer great flexibility in allowing the use of more than one materials at the same time makes it possible to fabricate components that is made of composite material as well as compositionally graded material [4, 5]. Laser metal deposition technology is still fairly new and a number of unresolved issues still exist. The key processing parameters in the LMD technology play important role in the and proper processing parameters control is key at reducing the cost of the manufacturing process such as the need to eliminate secondary finishing operations [6-10]. LMD process is a contactless manufacturing technology that makes it an important technology for processing difficult to machine materials. A number of studies have been reported in the literature that which relates the influence of processing parameters to the properties of the material produced using the LMD process [11-15]. The processing parameters in the LMD process were found to have great influence on the evolving properties of the deposited materials. In one of the authors' previous work, the influence of power density on the resulting surface roughness of Titanium alloy produced using LMD process was investigated. This study showed that, surface finish improves as the laser

Content from this work may be used under the terms of the Creative Commons Attribution 3.0 licence. Any further distribution of this work must maintain attribution to the author(s) and the title of the work, journal citation and DOI.
Published under licence by IOP Publishing Ltd

power density was increased. A number of industries are interested in titanium and its alloys due to the excellent properties of this alloy which include low density, high strength-to-weight ratio, and excellent corrosion resistance. such industries include the aerospace, chemical, petrochemical, marine and medical industries [16]. Ti6Al4V is an important Titanium alloy often termed the workhorse in the industry due to its vast application area [17]. Laser metal deposition process is an important alternative manufacturing process for difficult to machine material like titanium and its alloys. The laser metal deposition process parameters need to be properly controlled in order to control the properties of the deposited materials. To further build on the past study [18], this study investigates the influence of powder flow rate on the surface finish produced during the LMD process. Ti6Al4V was used in this study due to its excellent mechanical properties and bio-compatibility which make it useful in aerospace and biomedical industries [19].

2. Experimental Methods

The substrate used in this study is an annealed 5mm thick 99.6% pure Ti6Al4V sheet has supplied by VSMPO-AVISMA Corporation in Russia. Before the deposition process was commenced, the substrates prepared through sandblasting and then washed in acetone and dried. The reason for this preparation was to make the surface of the substrate to become less shining which is desirable for better laser absorption. The Ti6Al4V powder is also 99.6% pure and of particle size of between 120 and 350 μm. The laser that was used in this study is a 4.0 kW Nd: YAG laser attached to the end effector of a Kuka robot. The laser power, the scanning speed, and the gas flow rate were maintained at 3.0 kW, 0.004 m/s, and 2 l/min respectively. The laser spot size was also kept at a value of 2mm which is at a focal length of 195mm above the substrate. Powder flow rate was set between 1.44 g/min and 7.2 g/min. A glove box was provided and filled with argon gas to keep the oxygen at a low level and also as a powder carrying gas. The deposited samples needs to be prevented from environmental attach as it could also result in degrading the surface finish. The laser was used to create a melt pool on the substrate where the Ti6Al4V powder was delivered during the laser metal deposition process. The powder was introduced to the melt pool using coaxial powder delivery system that iv also attached to the end effector of the robot. The schematic diagram of the laser metal deposition process is shown in Figure 1.

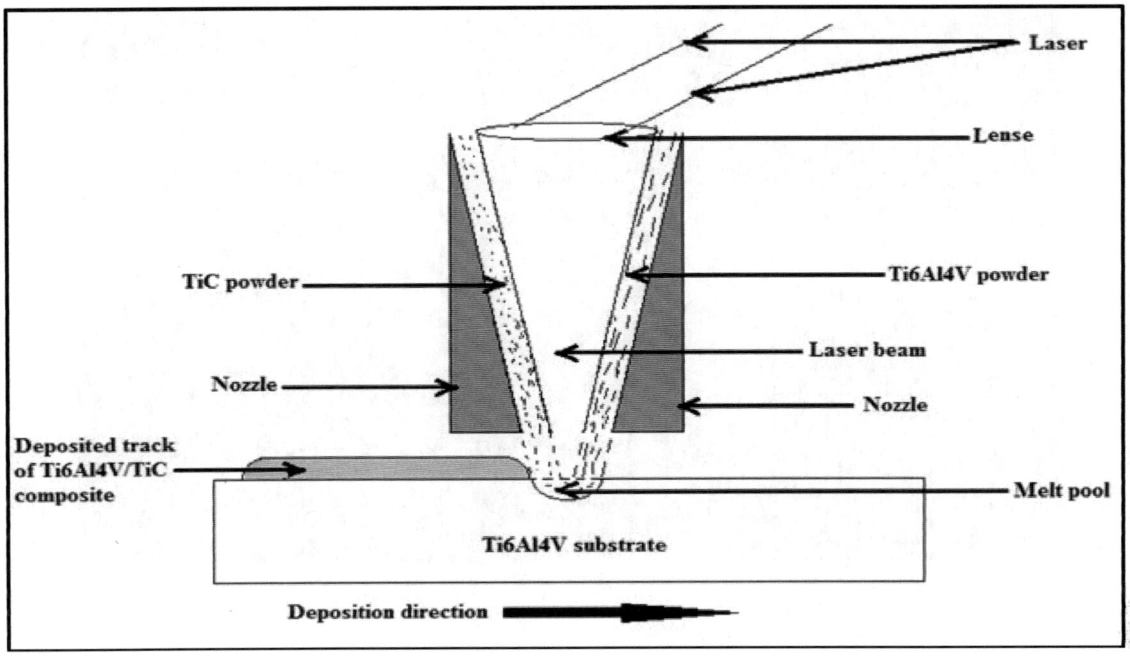

Figure 1: Schematic diagram of the laser metal deposition process [19]

After the deposition experiment was completed, the samples produced were thoroughly cleaned using acetone so as to remove any loose powder particles and dirt that is found attached to the samples that could cause error in the surface roughness measurements. After the samples were cleaned, the surface roughness measurement was carried out with the Jenoptik surface analyzer (see Figure 2) which is equipped with Hommelmap 6.2 software. The measurement was conducted using measuring condition according to 'BS EN ISO 4288:1998' standard [20]. The measuring conditions used are as follows: the effective measuring length is 4.8 mm, the cut-off length is 0.8 mm, sliding speed is 0.50 mm/s and the measuring range employ is 400 µm. The measurement was taken five times on each sample and the arithmetic mean of the surface roughness profiles (Ra) for each of the samples was recorded.

IOP Conf. Series: Materials Science and Engineering **391** (2018) 012005 doi:10.1088/1757-899X/391/1/012005

Figure 2 Photo of Jenoptik surface analyzer used in this study

3. Results and Discussions

Table 1 presents the average surface roughness at each set of processing parameters. Figure 3 shows the graph of the average surface roughness versus powder flow rate. The surface roughness was seen to increase as the powder flow rate was increased. At lower powder flow rate, the available laser power was able to efficiently melt the deposited powder which results in creation of bigger melt pool. Bigger melt pool takes longer to solidify and cool down which results in production of smoother surface finish. By increasing the powder flow rate, there is improper melting of the delivered powder since the laser power was fixed.. This results in rougher surface. Also, the melt pool that was created at higher powder flow rate was smaller due to improper melting of the powder which makes it to solidified more quickly thereby contributing to poor surface finish. The importance of this study helps to determine the proper combination of processing parameters so the desired surface finish can be achieved . This will in turn have impact on the economy of the entire process. With the use of proper processing parameters, Requirement for expensive secondary finishing operation will be eliminated when the LMD process is use to fabricate 3D product and in repair of parts.

Table 1. Results of average surface roughness

Sample No.	Powder flow rate (g/min)	Average surface roughness (μm)
PM	-	0.42
A	1.44	4.5
B	2.88	7.7
C	4.32	11.9
D	5.04	15.2
E	5.76	19.2
F	7.2	27.8

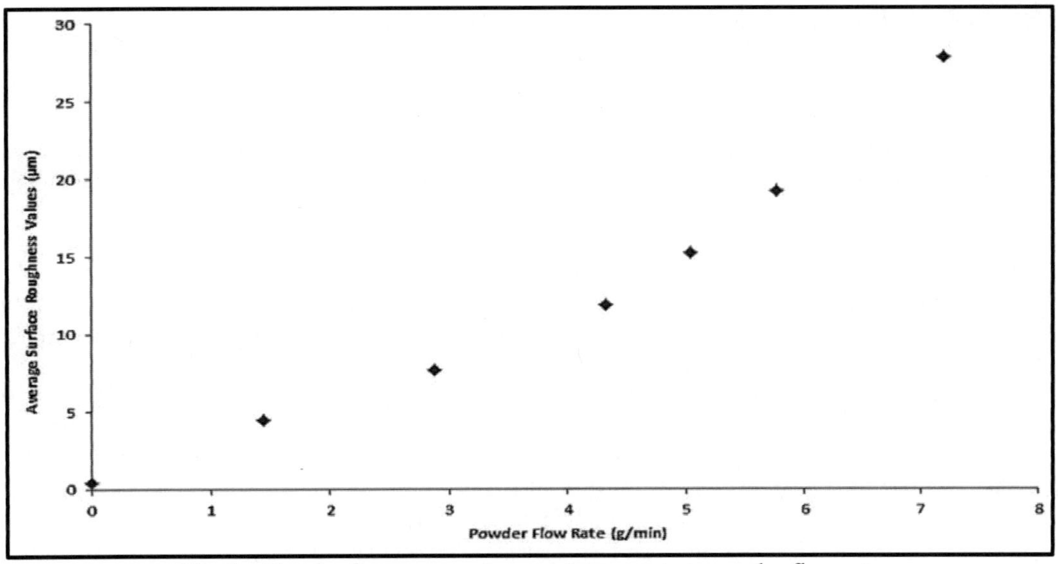

Fig. 3: Graph of average surface roughness against powder flow rate

4. Conclusions

This paper investigated the influence of powder flow rate, an important processing parameter in laser metal deposition process, on the quality of surface finish produced when titanium alloy, Ti6Al4V was produced with laser metal deposition process. The powder flow rate was varied between 1.44 to 7.2 g/min. Other processing parameters were kept constant through the deposition process. The results obtained in this research work revealed that, the powder flow rate has a significant influence on the quality of surface finish achievable in the laser metal deposition process. It was found that, the quality of surface finish improves with decreasing the powder flow rate. It can be concluded that the powder flow rate should be carefully controlled during the laser metal deposition process so as to reduce the need for secondary finishing operations which could be very expensive and affect the economy of the process. This study is important most especially in the aerospace industry, where highly complex parts are produced and also in product repair and remanufacturing process.

Acknowledgment

The authors appreciate the support received from the University of Johannesburg research council and also the Nation Laser Center-CSIR, Pretoria South Africa for the laser rental pool grant.

References

[1] Scott, J., Gupta, N., Wember, C., Newsom, S., Wohlers T. and Caffrey, T., "Additive Manufacturing: Status and Opportunity, "*Science and Technology Policy Institute.* Available from: https://www.ida.org/stpi/occasionalpapers/papers/AM3D_33012_Final.pdf [Accessed on 01 March 2017] (2012)

[2] Pinkerton, A. J., Wang, W. and Li, L., "Component repair using laser direct metal deposition, "*Proc. IMechE Part B: J. Engineering Manufacture*, 222, pp. 827-836. (2008).

[3] Graf, B., Gumenyuk, A. and Rethmeier, M., "Laser metal deposition as repair technology for stainless steel and Titanium alloys, "*Physics Procedia*, 39: pp. 376-381, (2012).

[4] Mahamood, R.M and Akinlabi, E.T., "Laser metal deposition of functionally graded Ti6Al4V/TiC, "*Materials & Design*, 84, pp. 402-410, (2015).

[5] Mahamood, R. M. and Akinlabi, E. T., "Functionally graded material" Springer Science Publisher, Switzerland. (2017)

[6] Mahamood, R.M and Akinlabi, E.T.," Effect of Scanning Speed and Powder Flow Rate on properties of LMD Titanium-alloy," International Journal of Mechanical engineering and technology , doi:10.1007/s00170-016-9954-9. (2017)

[7] Mahamood, R.M and Akinlabi, E.T.," Microstructure and Mechanical Behaviour of Laser Metal Deposition of Titanium Alloy, *"Lasers in Engineering*, 35(1-4). pp. 27-38, (2016)

[8] Mahamood, R.M and Akinlabi, E.T., "Process Parameters Optimization for Material Deposition Efficiency in Laser Metal Deposited Titanium Alloy," *Lasers in Manufacturing and Materials Processing*, 3(1), pp. 9-21,(2016) DOI 10.1007/s40516-015-

[9] Mahamood, R. M., Akinlabi, E. T., Shukla, M. and Pityana, S., "Material Efficiency of Laser Metal Deposited Ti6Al4V: Effect of Laser Power, *"Engineering Letters*, 21:1, EL_21_1_03, available online at http://www.engineeringletters.com/issues_v21/issue_1/EL_21_1_03.pdf (2013).

[10] Mahamood, R.M and Akinlabi, E.T., "Effect of Laser Power on Surface Finish during Laser Metal Deposition Process," WCECS 2014. 2, pp. 965-969, (2014).

[11] Zhong, C., Biermann, T., Gasser, A. and Poprawe, R., "Experimental study of effects of main process parameters on porosity, track geometry, deposition rate, and powder efficiency for high deposition rate laser metal deposition, Journal of Laser Application. 27 (2015) http://dx.doi.org/10.2351/1.4923335

[12] Campanelli, S. L., Angelastro, A., Signorile, C. G. and Casalino, G. "Investigation on direct laser powder deposition of 18 Ni (300) marage steel using mathematical model and experimental characterization, *"The International Journal of Advanced Manufacturing Technology*. pp. 1—11, (2016) doi="10.1007/s00170-016-9135-x",

[13] Mahamood, R. M., Akinlabi, E. T., Shukla, M. and Pityana, S., "Characterizing the Effect of Processing Parameters on the Porosity Properties of Laser Deposited Titanium Alloy, "*International Multi-conference of Engineering and Computer Science (IMECS 2014).*

[14] Mahamood, R. M., Akinlabi, E. T., Shukla, M. and Pityana, S., "Characterizing the Effect of Laser Power Density on Microstructure, Microhardness and Surface Finish of Laser Deposited Titanium Alloy," *Journal of Manufacturing Science and Engineering.* 135(6) (2013). doi:10.1115/1.4025737.

[15] Zhang, K., Zhang, X. M., and Liu, W. J., "Influences of Processing Parameters on Dilution Ratio of Laser Cladding Layer during Laser Metal Deposition Shaping," *Advanced Materials Research.* 549 785-789, (2012).

[16] Destefani, J. D. " Introduction to Titanium and Titanium Alloys," *Metals Handbook ASM International, Materials Park*, 2, pp. 586-591, (1998).

[17] Peters, M., Kumpfert, J., Ward, C.H. and Leyens, C. "Titanium alloys for aerospace applications, in: Titanium and Titanium alloys, "*Advanced Engineering Materials*, 5, 419- 427(2003).

[18] Makoto, K., Daishi, I., and Naoyuki, K., "Combustion synthesis of porous TiC/Ti by a self-propagating mode, "*Materials* 3,3939–3947, (2010).

[19] Mahamood, R. M., Akinlabi, E. T., Shukla, M. and Pityana, S., "Scanning Velocity Influence on Microstructure, Microhardness and Wear Resistance Performance on Laser Deposited Ti6Al4V/TiC Composite," *Materials and Design,* 50, pp. 656-666, (2013).

[20] BS EN ISO 4288:1998, Geometric product specification (GPS). Surface texture. Profile method: Rules and procedures for the assessment of surface texture, BSI.

Enhanced Recycling of Waste Copper Smelter Dust from Existing and Proposed Smelter Plant at the Palabora Copper (pty) ltd, impopo, South Africa

[1]Daniel O. Okanigbe*, [1]Abimbola P.I. Popoola* and [2]Abraham A. Adeleke

[1]Department of Chemical, Metallurgical and Materials Engineering, Tshwane University of Technology, Pretoria, 0183, South Africa
[2]Department of Materials Science and Engineering, Obafemi Awolowo University, Ile-Ife, 220002, Nigeria
*Corresponding Authors:okanigbedo@tut.ac.za;PopoolaAPI@tut.ac.za; odanielifeoma@yahoo.com,
Tel: +27 12 382 3513, +27634243699, +27618851872; Fax: +27 12 382 3532.

Abstract. The whole facilities at the Palabora Copper (PTY) Ltd (PC), Limpopo, South Africa (i.e. both existing and new) are under obligation to comply with the new plant MES by the 1st of April, 2020. At the moment all processes on the PC mine conform to the MES for existing plant with the exception of the copper smelter and the associated sulphuric acid plant. Despite considerable investment by PC over the last few years in equipment and maintenance, due to the short comings of the copper smelting technology employed on site, certain relevant MES have proven to currently be unachievable. The generation of the waste CSD is considered a global issue and at PC the direct smelting process option is the recycling method currently being used to manage the generated waste CSD at the copper smelting plant. This recycling method is neither cost effective nor energy efficient and it leads to the production of worthless residues. As a result, an enhanced hydrometallurgical based recycling process option is recommended as an appropriate recycling process option. Hence, it was concluded that an industrial scale centrifugal separator like the Knelson or Falcon concentrators be stationed in the existing and proposed plants as a separate but integral part of the ore processing system, so as to reduced/removed the reactive materials it often contains before subjecting it to a hydrometallurgical and Nano technological processing, in order to recover the copper value in it as precipitates of nanoparticles.

Keywords: direct smelting, recycling, waste CSD, hydrometallurgical processing, nano-technological processing

1. Introduction

Palabora Copper (PTY) Ltd (PC), Limpopo, South Africa was issued a provisional Atmospheric Emissions Licence (AEL) with respect to the national environmental management air quality, Acct 39 of 2004 (NEM: AQA) provisional atmospheric emissions license (NEM: AQA) by the Limpopo, provincial department of economicsdevelopment, environment and tourism (LEDET) in April 2015 and is required to meet the minimum emission standards (MES) contained in government notice (GN) R893, promulgated in November 2013in terms of section 21of the NEM: AQA. GN R893 prescribes MES for both existing and new facilities. All facilities (i.e. both existing and new) are required to comply with the new plant MES by

Content from this work may be used under the terms of the Creative Commons Attribution 3.0 licence. Any further distribution of this work must maintain attribution to the author(s) and the title of the work, journal citation and DOI.
Published under licence by IOP Publishing Ltd

1st of April, 2020. Currently all processes on the PC mine comply with the MES for existing plant with the exception of the copper smelter and the associated sulphuric acid plant. Despite substantial investment by PC over the last few years in equipment and maintenance, due to the limitations of the out dated copper smelting technology employed on site, certain relevant MES have proven to currently be unachievable. Even though PC is committed to managing its impact on air quality and in order to achieve compliance with relevant MES for new plant at the copper smelter and sulphuric acid plant before 01 April,2020, PC is proposing to implement the smelter Retrofit project. This proposed smelter retrofit project will involves the replacement of the furnace, which will consist amongst other activities a double side blown (DSB) feeding and dust collection system. Howbeit, the actual method to be used in the management of this waste CSD in the dust collecting system of the new plant was not specified in aforementioned report, which invariably leads back to statuesque; the recycling via direct smelting of the dust.
However, the direct smelting process option requires investment in expensive equipment which results in high energy consumption, thus leading to the production of worthless residues [2, 3, 10, 11 and 16]. As a result, an enhanced hydrometallurgical based recycling process option is preferred [12]. In addition, the leach-solvent extraction-electrowinning processing option has been the conventional route of recovering copper from its source as copper cathode slabs [1, 7]. But, recovering the copper value in this waste CSD as precipitates of nano-particles is to be preferred more, owing to vast applications in areas such as heat transfer and microelectronics together with its energy efficiency advantage when compared to electrowinning. A study that has been carried out at laboratory scale [4]; In this instance the authors combined hydrometallurgy with nanotechnology to recover the contained copper in the waste CSD from Iran as nanoparticles of copper oxide (Tenorite). This approach thus illuminates the issues and solutions surrounding the management of the waste CSD from PC.
Hence, this report will start by presenting an overview on the recycling of metallurgical wastes, thereafter the focus will shift to, firstly, ore processing at the existing Palabora copper (PTY) Ltd plant, secondly, ore processing at the proposed Palabora copper (PTY) Ltd plant and finally the proposed enhanced recycling method for the waste CSD at both the existing and proposed Palabora copper (PTY) Ltd plant.

2. An Overview on Recycling and the Waste Copper Smelter Dust

At this moment in time, the disposal of metallurgical wastes such as the waste copper smelter dust (CSD) is becoming costly as a result of increasingly severe environmental code of best practice [8]; in addition, the physical and chemical nature of the waste CSD brands it "hazardous waste" [13]. While in Africa, the South African National Standard (SANS) annual target level of dust is 300 $mg/m^2/day$ and the industrial action level of dust are 1200 $mg/m^2/day$ [14]. Added to this, the national environmental management waste act (Act 59 of 2008) and subsequently the national waste management strategy (NWMS 201) are based on the principles of the waste management hierarchy (Figure 1), which is a framework used worldwide. Recycling in this hierarchy (Figure 1) is one of the important waste management options in order to prevent disposal to the landfill. Despite the recycling industry being in its infancy stage in South Africa, with a need for further development. Hence, emphasis is therefore placed on promoting recycling of waste through the creation of an enabling environment for industry to develop processing facilities as well as educating and assisting other role players, especially the waste producers to also facilitate recycling.

There is one of two basic options to the management of wastes once generated; it is either recycled or disposed. Recycling represents actions that manage in a way, thus resulting in the use of the materials, making it serve as feed for some product or an application distinct from disposal. This is done by segregating waste at source or by extracting materials from it by various techniques, physical, chemical, thermal or biological. Metals and their alloys (steel, aluminum, copper, nickel etc.) are traditionally recycled whereas minerals and products derived from them (e.g. fertilizer, cement) are not recycled. There is growing concern to efficiently manage minerals and the products generated in recovering them from ore bodies. These include waste rocks, tailings, and slags, fly ash and dross.

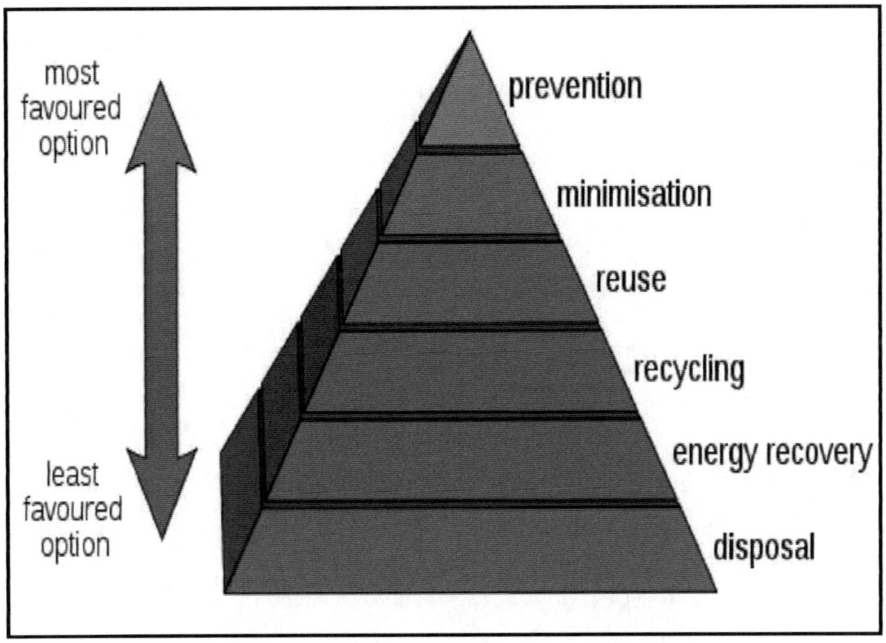

Figure 1: Image showing the waste management hierarchy [5]

Most metals are not consumed (though some of them are subjected to corrosion) and recycling has great environmental and economic pay off.Concentrated scraps, diluted streams and effluent, spent catalysts, arc furnace dust, plating solutions, and process water are some of the examples. Furthermore, recycling is emphasized as it potentially prevents useful material from being combusted or land filled, thus reducing the cost of waste disposal and reducing landfills and tailing ponds. In the metal industry recycling is one of the most effective ways of reducing energy requirement. As metal recycling is done by treating the metal or engineering the material ore, the energy associated with mining the ore is eliminated.

However, waste minimization or ideally elimination of waste is a much sought after objective. What's more technological advances in recent years, together with scientific understanding of the processes that occur, have led to new development towards achieving this goal. It has to be recognized, nevertheless, that recycling continues to be and will remain so for many years to come a major occupation in metallurgical industry for two main reasons. Firstly, metals are a finite source; in industrialized countries, virgin sources of metals natural ores are steadily getting depleted. Recycling of metals from scrap, discarded metal products and engineered materials is an obvious necessity to ensure availability of the metals in desired quantities. Secondly, in many metallurgical processes, nature of reactions is such that products of no direct value for the primary metal industry are inevitably generated. Often considered to be "waste" many of them find their way into landfills or tailings ponds. Furthermore, such waste products have accumulated in huge quantities and call for serious action for resource recovery and environmental reclamation. Resource recovery from such "waste" or converting them to useful by-products is a growing necessity. They will be increasingly in demand to achieve the goal of sustainable development, constant industrial growth for healthy living standards, which requires clean environment and uninterrupted availability of resources.

3. Ore Processing at Existing Palabora Copper (PTY) LTD Plant (EP)

The existing copper smelter is a metallurgical sulphide and converting facility, commissioned in 1966 and comprising of the following:

- A reverberatory furnace (Figure 2A) with dimensions of 36 by 10 m, serving as asmelting furnace with peripheral facility.
- Three Pierce smith converting copper matte into blister copper.
- Three anode furnaces for casting copper into anode for electro refining.
- A sulphuric acid plant with a maximum off-gas treatment capacity of 3,000 normal cubic meters(Nm^3) per minute which fixes SO^2 gas emitted as a waste gas from the converting process, into sulphuric acid.

3.1 Reverberatory Furnace in the EP

Copper concentrate from the mineral processing stage is smelted in the reverberatory (Fig. 2A), converter and anode furnaces, the sulphur content of the concentrate combines with oxygen to release as SO_2, thus purifying the copper. The sulphur to SO_2 ratio is 1: 2 and approximately 75% of the sulphur contained in the copper concentrate is captured during smelting process.

3.2 Electrostatic Precipitator in the EP

Off-gas from the reverberatory furnace (Fig. 2B) passes through two waste heat boilers and a balloon flue before it reaches a final process of cleaning by electrostatic dust precipitation.The flue dust is returned to the furnace and the low concentration off-gas can be discharged directly into the atmosphere via a 152 meter high concrete stack. It can also be treated in the wet gas scrubbing plant, if the precipitator is off-line. In this instance, once scrubbed, the off-gas is released to the atmosphere via a 70 meter high clean gas stack.The different size ranges for various types of dust, smoke, and fog particle matters and working ranges for different types of gas cleaning equipments are presented on Figure 3 and 4.

3.3 Existing Management Practice of Generated Waste CSD

Over the years, PC has embarked on several studies to identify options to sustain and improve environmental performance. Currently all processes on the mine comply with the MES for the existing plant, with the exception of the copper smelter and the associated sulphuric acid plant. Despite substantial investment by PC over the last few years in equipment and maintenance, certain relevant MES have proven to currently be unachievable based on but not limited to the integrated nature of PC's activities, the outdated reverberatory furnace technology, incompatible with the acid plant and associated limited efficiency of the single contact acid plant.

Currently the management practice of waste CSD generated during the smelting of the copper concentrate (Fig. 2B) at the PC is recycling it together with new concentrates as feed. An approach that has been condemned [8]because it requires investment in expensive equipment, results in high energy consumption and thus, leads to the production of worthless residues [2, 3, 10, 11 and 16],and because of its ultra-fine nature (Fig. 3) it causes damage to the refractory bricks of the reverberatory furnace [12].

Figure 2: Images of (A) PC's existing reverberatoy furnace and (B) off-gas from the reverberatory furnace [6].

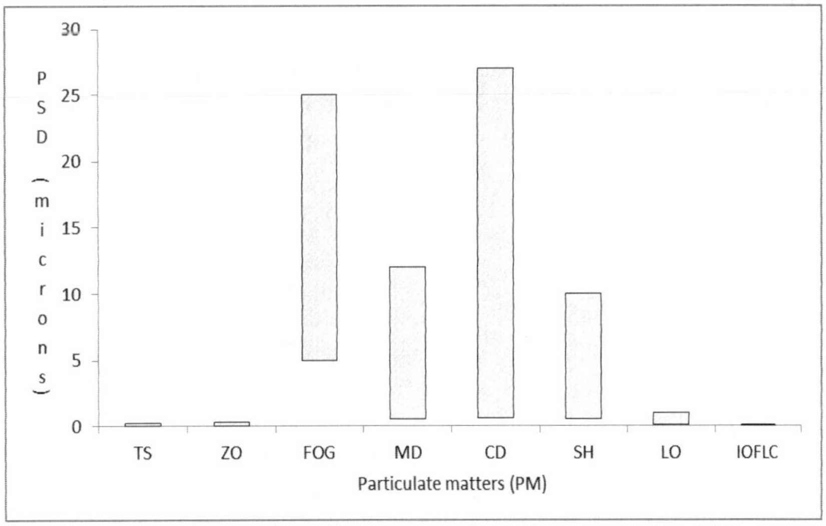

Key: PSD = Particle size distribution; TS = Tobacco smoke; ZO = Zinc Oxide; MD = metallurgical dust; SH = SO_3-H_2SO_4;
LO = Lead oxide; IOFLC = iron oxide from LD converter; CD = cement dust.

Figure 3: Size range for various types of dust, smoke, and fog [14]

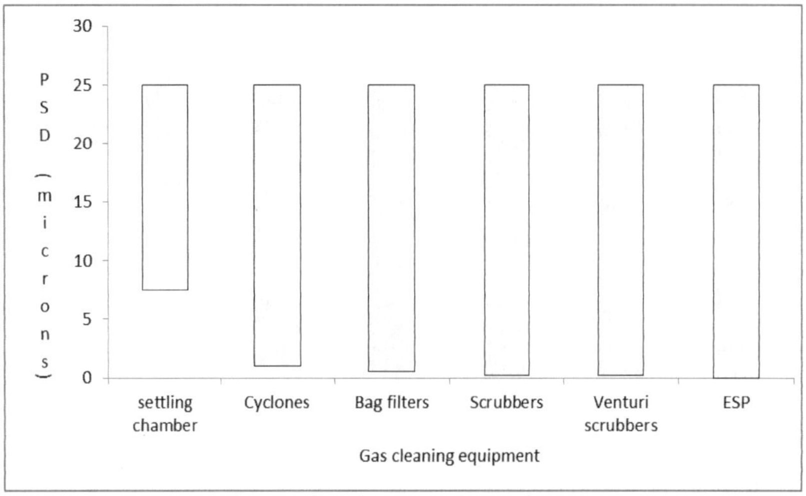

Key: PSD = Particle size distribution; ESP = electrostatic precipitator

Figure 4: Working ranges for different types of gas cleaning equipment [14]

4. Ore Processing at the Proposed Palabora Copper (PTY) ltd Plant

In the proposed new plant (NP) Both the draft basic assessment and environmental management programme report dated October, 2016 and titled: Proposed Smelter Retrofit Project at Palabora Copper was submitted for environmental authorizations in terms of the national environmental management act, 1998 and the national environmental management waste act, 2008 in respect of listed activities that have been triggered by applications in terms of the mineral and petroleum resources development act, 2002 (MPRDA).

5. Proposed Enhanced Recycling of the Waste csd at the Existing and Proposed Palabora Copper (pty) Ltd Plant

A robust data from the physical, chemical, mineralogical and surface morphological characterization of PC's waste CSD has been reported in the open literature [9]. Therein, the authors recommended a mineral processing stage be included in the conventional hydrometallurgical based process options of roast –leach or direct leaching so as topositively impact on the overall processing economy and environment.

6. Conclusion

Based the research conducted and preliminary results presented, it is evident that there is a major gap in knowledge with respect to a mineral processing route and considering the ultra-fine nature (below 53 micron) of the waste CSD, the valuable metal(s) in it can be concentrated using an industrial scale centrifugal separator like the Knelson or Falcon concentrators stationed in the existing and proposed plants as a separate but integral part of the ore processing system. The concentrate which should have reduced/removed reactive materials can subsequently be subjected to hydrometallurgy processing, in order to take its contained copper into solution and further adding value to it by recovering it as precipitates of nano-particles.

Acknowledgement

The authors would wish to thank the Council for scientific and industrial research (CSIR), Pretoria, South Africa andTshwane University of Technology (TUT) for their financial support in the course of this project, while also appreciating PC for providing the waste CSD used for this study.

References

[1] Alguacil, F.J., Garcia-Diaz, I., Lopez, F. and Rodriguez, O., 2015. Recycling of copper flue dust via leaching-solvent extraction processing. Desalination and Water Treatment, 56(5), pp.1202-1207.

[2] Bakhtiari, F., Atashi, H., Zivdar, M. and Bagheri, S.S., 2008. Continuous copper recovery from a smelter's dust in stirred tank reactors. International Journal of Mineral Processing, 86(1), pp.50-57.

[3] Bakhtiari, F., Zivdar, M., Atashi, H. and Bagheri, S.S., 2008. Bioleaching of copper from smelter dust in a series of airlift bioreactors. Hydrometallurgy, 90(1), pp.40-45.

[4] Darezereshki, E. and Bakhtiari, F., 2013. Synthesis and characterization of tenorite (CuO) nanoparticles from smelting furnace dust (SFD). Journal of Mining and Metallurgy B: Metallurgy, 49(1), pp.21-26.

[5] https://en.wikipedia.org/wiki/Waste_hierarchy

[6] http://www.palabora.com/smelter.asp

[7] Kordosky, G.A., 2002. Copper recovery using leach/solvent extraction/electrowinning technology: Forty years of innovation, 2.2 million tonnes of copper annually. Journal of the South African Institute of Mining and Metallurgy, 102(8), pp.445-450.

[8] Montenegro, V., Sano, H. and Fujisawa, T., 2008. Recirculation of Chilean copper smelting dust with high arsenic content to the smelting process. Materials transactions, 49(9), pp.2112-2118.

[9] Okanigbe, D.O., Popoola, A.P.I. and Adeleke, A.A., 2016. Characterization of Copper Smelter Dust for Copper Recovery. Procedia Manufacturing, 7, pp.121-126.

[10] Pinto, I.S. and Soares, H.M., 2012. Selective leaching of molybdenum from spent hydrodesulphurisation catalysts using ultrasound and microwave methods. Hydrometallurgy, 129, pp.19-25.

[11] Pinto, I.S. and Soares, H.M., 2013. Recovery of molybdates from an alkaline leachate of spent hydrodesulphurisation catalyst–proposal of a nearly-closed process. Journal of Cleaner Production, 52, pp.481-487.

[12] Qiang, L., Pinto, I.S. and Youcai, Z., 2014. Sequential stepwise recovery of selected metals from flue dusts of secondary copper smelting. Journal of Cleaner Production, 84, pp.663-670.

[13] Regulation, E.C., 1999. No 1907/2006 of the European Parliament and of the Council of 18 December 2006, concerning the Registration. Evaluation, Authorisation and Restriction of Chemicals (REACH), establishing a European Chemicals Agency, amending Directive, 45, pp.1-849.

[14] Rosenqvist, T., 2004. Principles of extractive metallurgy. Tapir academic press.

[15] SANS. 2005. South African National Standard: Ambient air quality–Limits for common pollutants.

[16] Wu, J.Y., Chang, F.C., Wang, H.P., Tsai, M.J., Ko, C.H. and Chen, C.C., 2015. Selective leaching process for the recovery of copper and zinc oxide from copper-containing dust. *Environmental technology*, 36(23), pp.2952-2958.

IOP Conf. Series: Materials Science and Engineering **391** (2018) 012007 doi:10.1088/1757-899X/391/1/012007

Effects of Quenchants on Impact Strength of Single-Vee Butt Welded Joint of Mild Steel

O. M. Ikumapayi[*1,3], I. P. Okokpujie[2], S. A. Afolalu[2], O. O. Ajayi[2], E. T. Akilabi[1], O. P Bodunde[3]

[1]Mechanical Engineering Science Department, University of Johannesburg, South Africa
[2]Mechanical Engineering Department, Covenant University, Ota, Nigeria
[3]Mechanical and Mechatronics Engineering Department, Afe Babalola University, Ado Ekiti
*Corresponding author. Tel: +2348036049538, +27838098288, e-mail: imhade.okokpujie@covenantuniversity.edu.ng, oikumapayi@uj.ac.za

Abstract- The effects of quenching medium on the impact behaviour of mild steel welded joints were investigated, single – vee butt welded joint was employed in welding the samples. The welding was carried out at 100A and at a terminal voltage of 140V (14kW), using gauge 8 coated electrode (4mm size) throughout the welding processes. Four (4) quenching media were employed which are Brine solution (0.1 Molar concentration), two (2) Litres of water, two (2) Litres of diesel oil and abundant air. 8mm thick mild steel bars were used for the heat treatment process at variable annealed temperature steps of 200^0C – 400^0C – 600^0C. The Izod impact machine was employed in carrying out the Impact Test. The results obtained from the test carried out show that quenching in brine solution gives least average impact strength while normalized in air gives highest average impact strength and is the mildest of all the quenchants. The overall results show that air was the best medium for quenching welded mild steel followed by diesel oil, then water and lastly brine solution for quenching welded parts or components under influence of impact or sudden loading. The results will enable engineers, welding personnel and roadside welders select the best and most economic quenchant that will be best suitable for quenching welded components or parts made of mild steel subjected to impact or shocking loading.
Keywords: Quenchants, Mild Steel, Single-Vee, Butt welded Joint, Impact strength, annealed

1. Introduction

Many welding processes are accomplished by heat alone, with no pressure applied; others by a combination of heat and pressure; and still others by pressure alone, with no external heat supplied. In some welding processes a filler material is added to facilitate coalescence. The assemblage of parts that are joined by welding is called a weldment. Welding is most commonly associated with metal parts, but the process is also used for joining plastics [1].
Although in its present form it has been used since about the beginning of 20th century but it is fast replacing other joining processes like riveting and bolting. At times it may be used as an alternative to casting [2].
Presently welding is used extensively for fabrication of vastly different components including critical structures like boilers and pressure vessels, ships, off-shore structures, bridges, storage tanks and spheres, pipelines, railways coaches, anchor chains, missile and rocket parts, nuclear reactors, fertilizer and chemical plants, structural, earth moving equipment, plate and box girders, automobile bodies, press frames and water turbines. Welding is also used in heavy plate fabrication industries, pipe and tube fabrication, joining drill bits to their shanks, automobile axles to brake drums, lead wire connections to transistors and diodes, sealing of containers of explosives like, nitroglycerine, welding of cluster gears, and the like [2].

Content from this work may be used under the terms of the Creative Commons Attribution 3.0 licence. Any further distribution of this work must maintain attribution to the author(s) and the title of the work, journal citation and DOI.
Published under licence by IOP Publishing Ltd

Steel is an important engineering material. It has found applications in many areas such as vehicle parts, truck bed floors, automobile doors, domestic appliances, Ship's hull, structural castings, Railway rolling stock, automotive castings, Hot metal ladles, Rolling mill equipment, Rolls and rollers, Machines and tools, Mine and quarry equipment, Oil and petroleum equipment etc. It is capable of presenting economically a very wide range of mechanical and other properties [3]. For instance, the most commonly used material for petroleum pipelines is mild steel, this is because of its strength, ductility, weldability formability and its amenability to heat treatment for varying mechanical properties [4, 5].

Steel is an effective, cheap and commercially available material for structural applications; large tonnages of commercial steels in Nigeria contains very little alloying elements with carbon composition at varied percentage and thus are classified as low carbon, medium carbon and high carbon steel, [6]. Due to its amenability to heat treatment processing and alloying, this segment of steel finds its wide application in both structural and domestic applications [7]. These processes are widely utilized to achieve high mechanical (high yield strength, high proportional limit, and high fatigue strength) properties for their respective applications. The desirable properties of medium carbon steel can be achieved by adding suitable alloying elements and secondly by various conventional heat treatment [8]. The mechanical strength of medium carbon steels can also be improved by quenching in appropriate medium. However, the major influencing factors in the choice of the quenching medium are the kind of heat treatment, composition of the steel, the sizes and shapes of the parts [9, 10]. These properties can further be enhanced by welding.

Susheel and Syed [11] considered welding parameters such as voltage, current and welding speed on depth of penetration. The material was carbon alloy steel(0.14%C) of dimension 75 x 50 x6mm,voltage 18v and current was 250amps.The maximum penetration takes place with minimum speed even by maintain constant voltage and current. Samer [12] studied the influence of Welding Parameters on Mild Steel, this research investigated variation of depth of penetration, welding speed and heat input during the welding process. It was observed that depth of penetration is increases at each case of voltages and current with the increased heat input. Sunar et al. [13] worked on thermal and stress analysis of a sheet metal in welding, the study centered on a cantilever assembly subjected to heating at its fixed end which resembles the welding of a sheet metal, stress fields and temperature are computed during the heating process. The study employed finite element approach to predict stress fields. It is found that the temperature distribution in the transverse direction does not vary considerably but varies significantly in the longitudinal direction. The temporal change of temperature gradient induces stresses in the substrate material. However, the maximum magnitude of the Von Mises stress is less than the yield strength of the substrate material.

Talabi et al. [3] investigated the effect of welding variables on mechanical properties of low carbon steel welded joint. The study concentrated on the effect of welding variables on the mechanical properties of welded 10 mm thick low carbon steel plate, welded using the Shielded Metal Arc Welding (SMAW) method. Welding current, arc voltage, welding speed and electrode diameter were the investigated welding parameters. The welded samples were cut and machined to standard configurations for tensile, impact toughness, and hardness tests. The results showed that the selected welding parameters had significant effects on the mechanical properties of the welded samples. Increases in the arc voltage and welding current resulted in increased hardness and decrease in yield strength, tensile strength and impact toughness. Increasing the welding speed from 40-66.67 mm/min caused an increase in the hardness characteristic of the welded samples.

Singh et al. [14] worked on the effects of joint geometries on welding of mild steel by Shielded Metal Arc Welding (SMAW). This research compared the effects of variations in geometry of butt-joint welding on the mechanical properties of mild steel plate. The welding was carried out on different butt-joint designs, such as, square butt-joint, single V-joint, double V-joint and single J-joint, keeping all other process parameters like current, voltage, welding speed etc. as constant. The mechanical test and the microstructural investigation were carried out to analyse the change in mechanical and microstructural behaviour of the weld metal. The results of tests peformed revealed that the Double-V joint was the superior of all other joints, having better mechanical properties than other joints. Single-V was also up to the mark but the more width of HAZ was recorded in this case as compared to others.

Bodunde and Momohjimoh [15] studied effects of welding parameters on the mechanical properties of welded Low-Carbon Steel using two welding processes of Oxy-Acetylene Welding (OAW) and Shielded Metal Arc Welding (SMAW). Two different edge preparations on a specific size, 10-mm thick low-carbon steel, with the following welding parameters: dual welding voltage of 100 V and 220 V, various welding currents at 100, 120, and 150 Amperes and different mild steel electrode gauges of 10 and 12 were used. The tensile strength, hardness and impact strength of the welded joint were carried out and it was discovered that the tensile strength and hardness reduce with the increase in heat input into the weld. However, the impact strength of the weldment increases with the increase in heat input.

2. Materials and Methods

2.1 Materials

Mild steel samples were used in this work. The steels were obtained as 20mm by 12mm from commercially available mild steel plain bars at Ibadan gate Iron and Steel Market, Oyo State, Nigeria. Coated gauge 8 (Size 4mm) AWS E6013 (150 – 190Amp) Mild steel electrodes were used for the welding, this was chosen based on the thickness of mild steels plain bars used.
Four different quenchants were employed in this study namely, Brine solution (0.1 Molar concentration), two (2) Litres of water, two (2) Litres of diesel oil and abundant air.
The migatronic LTC 140 arc welding machine was used. The machine has a voltage terminal of 140V with a variable of 0 to 140A. Honnsfield balance impact machine was used for impact testing. The impact energy absorbed can be varied from between 0 and 50 ft.pd. Izod impact test was carried out on the specimen.

2.1.1 Material Compositions

A spark test analysis was carried out on the Mild Steel sample to know its composition. Table 1 shows the chemical analysis of as-received mild steel (medium carbon steel) containing 0.33% carbon content as carried out at the universal steel (U-Steel) Ltd, Ikeja, Lagos

Table 1: Spectrometer analysis result of as-received medium carbon steel (MCS) - Mild Steel

Element	C	Si	S	P	Mn	Ni	Cr	Mo	V	Cu	W	As	Sn	Co	Al	Ca	Zn	Fe
Percentage	0.3300	0.1740	0.0499	0.0341	0.8225	0.0911	0.0585	0.0018	0.0029	0.3031	0.0003	0.0060	0.0230	0.0094	0.0019	0.0002	0.0037	Bal

2.2 Methods

2.3 Welding Procedure

5cm long pieces were cut from the mild steel sample and one end of each piece was milled at 30^0 one side and joined to form single – vee butt joints and to ensure proper weld penetration as shown in figure 1 and Figure 2 represents model of a welded sample under investigation.

IOP Conf. Series: Materials Science and Engineering **391** (2018) 012007 doi:10.1088/1757-899X/391/1/012007

Gauge 8 Electrode (4mm size) , 140V and 100A (14kW) were used in this work, the same voltage and current setting were used for all the specimens to have a base of assessing the effect of quenching on the impact behaviour of mild steel welded joints. The samples were welded together in pairs to give a single – Vee butt joint.

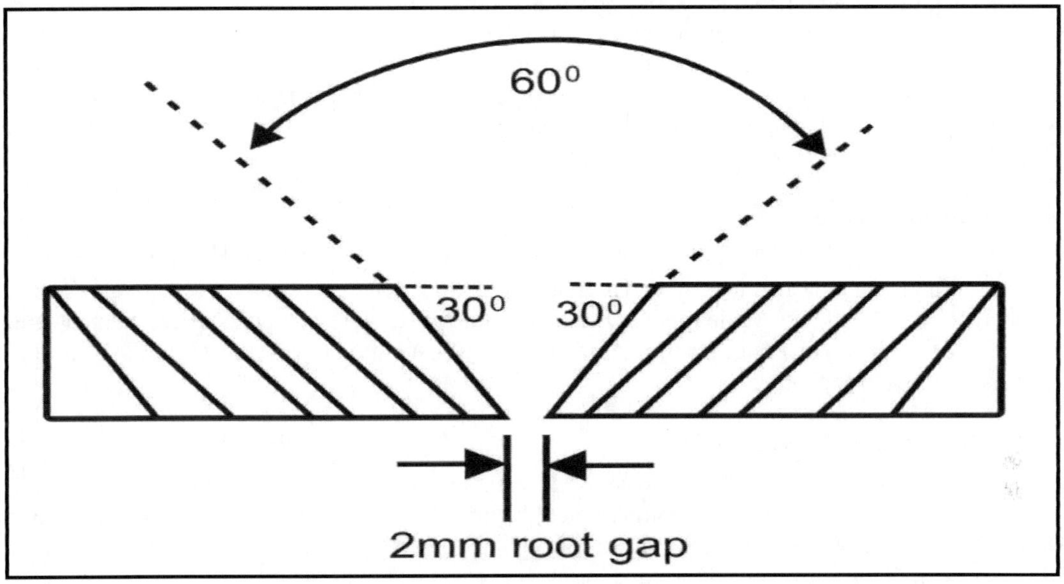

Figure 1: A pair of the milled piece ready for welding

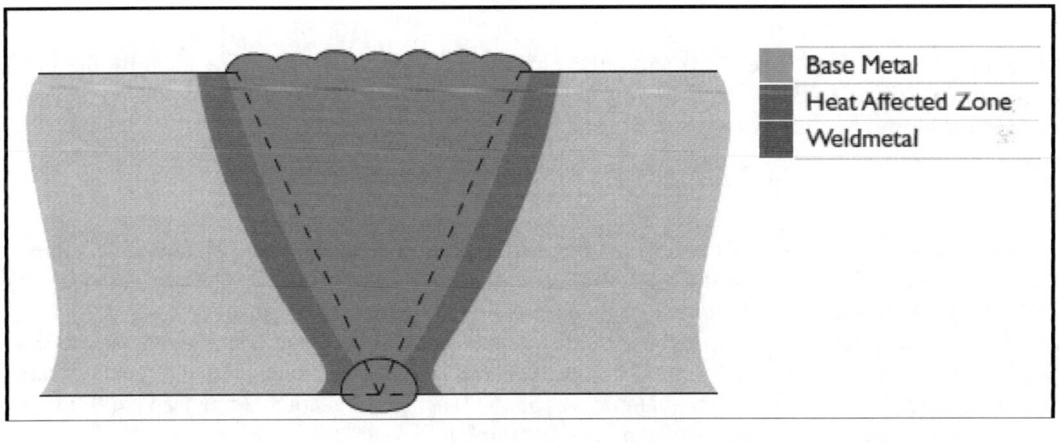

Figure 2: A model of a welded sample

2.4 Heat Treatment
2.4.1 Heat Treatment Process

A minimum of two pairs of welded samples were selected per treatment. Some of the conventional heat treatments procedure chosen include: annealing, normalizing, quenching (in diesel oil, Brine Solution and water)

IOP Conf. Series: Materials Science and Engineering **391** (2018) 012007 doi:10.1088/1757-899X/391/1/012007

2.4.2 Annealing process

After the machining to impact test specimens as shown in figure 3 and 4 then the impact test specimens were annealed in the electric furnace at different temperature range of $200^{\circ}C$, $400^{\circ}C$ and $600^{\circ}C$ respectively for each quenching medium before the impact test was conducted.
It is held at these temperatures for sufficient time (about 1 hour) to ensure thorough homogeneity and for all the material to transform into austenite. It is then cooled slowly inside the furnace to room temperature after the furnace has been turned off. The specimen was taken out of the furnace after reasonable hours of gradual loss of heat when the furnace temperature would have attained the nominal room temperature. The grain structure has coarse pearlite with ferrite or cementite.

2.4.3 Normalizing process

Immediately after set of machined samples, each samples of the medium carbon steel to be normalized were placed in the furnace and heated to temperature range of $200^{\circ}C$, $400^{\circ}C$ and $600^{\circ}C$ respectively. Two (2) pairs of 60^0 single – vee butt welded joints were normalized in air. The samples were retained at this temperature for the period of 1 hour for full transformation to austenite. They were later removed from the furnace and left in air for cooling.

2.4.4 Quenching Procedure

The selected samples for quenching were heated at different temperature range of $200^{\circ}C$, $400^{\circ}C$ and $600^{\circ}C$ respectively. At these temperatures the samples were held for 1 hour to ensure uniform homogeneity. In order to enhance the hardness, the red hot steel is directly and rapidly cooled

i. 2 pairs of 60^0 single – vee butt welded joints were quenched in diesel oil
ii. 2 pairs of 60^0 single – vee butt welded joints were quenched in water
iii. 2 pairs of 60^0 single – vee butt welded joints were quenched in brine solution (10 percent solution)

2.5 Material Testing

After the successful heat treatment operation, the various heat treated samples were taken for the Impact test. The test was performed on Izod Impact Testing Machine. Impact Test was conducted at various quenching medium.

2.5.1 The Impact Testing Procedure

Test specimens with a V notch in the fusion zone for impact test fracture were performed according to ASTM E23 standard. The impact test used in this research work was Izod test. The test was carried out on specially prepared specimens in the form of a notched bar (see figure 3 and figure 4).
For the purpose of this research, 8mm thick mild steel bars were used for the heat treatment process at variable annealed temperature of 200^0C, 400^0C and 600^0C respectively, and the Izod impact test was carried out on the sample pieces. The bar of the material was 55mm long, and a V-notch 2mm deep of 60^0 included angle was made at the centre of the specimens as shown in figure 3

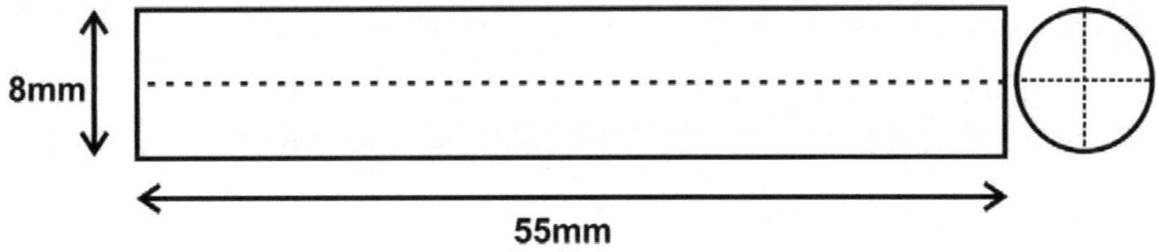

Figure 3: Unnotched impact specimen

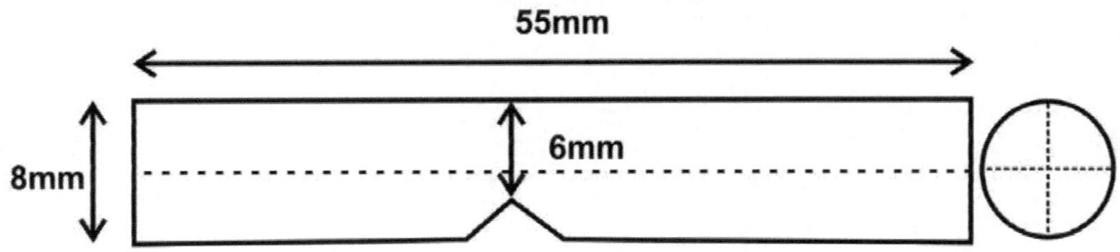

Figure 4: 60^0 V-notched

The prepared specimen was placed on the Anvil with V-notch gauge. The testing machine consists essentially of a weighted pendulum hammer which was then raised to a predetermined operating position in which it stores up 162.725J of potential energy at a velocity of 3.8m/s its weight and height above the point where it strikes the specimens and it was then released from a cooked position at a fixed height. Upon released, a knife edge mounted on the pendulum strikes and bends or fractures the specimen at the notch, which act as a point of stress concentration for this high velocity impact blow. The V-notch was located facing the striker and with the base of the V exactly in line with the top edge of the vice and was lined up with a small hang jig. The striker hits the specimens at the striking height 6mm above the V notch.

The height to which it swings over depends upon how much energy has been absorbed in fracturing or bending the specimen. If no energy were taken it would swing to its original height, while if all its energy were taken it would be arrested at the bottom of its swing. The amount of energy absorbed was shown by a pointer on the indicating dial and was recorded. The notches in the specimen are milled with a special form cutter or angle files. The Izod value was expressed in Joules (Nm) and represents the energy necessary to fracture or bend the test piece.

3. Results and Discussion

3.1 Results

Table 2, 3, 4 and 5 show the results obtained from the impact test for different quenching media i.e Brine Solution, water, diesel oil and air respectively, after annealed at different desired temperature steps of $200^0C - 400^0C - 600^0C$. The impact strengths were calculated for each medium at different temperature using the equation (1),

$$Impact\ Strength = \frac{E}{A} \quad (1)$$

E = Energy required to bend or fracture specimen

A = Cross-sectional area of the specimen as shown in equation (2)

Note: The diameter of each impact specimen used in this work is 8mm

$$A = \pi\left(\frac{d}{2}\right)^2 = \left(\frac{8}{2}\right)^2\pi = 50.27mm^2 = 50.27\ x\ 10^{-3}m^2 \quad (2)$$

Table 2: Impact value for annealed specimen quenched in Brine Solution

S/N	TEMPERATURE ($°C$)	IMPACT ENERGY ABSORBED (E) Ft - Ib_f	Joules	IMPACT STRENGTH $=\left(\frac{E}{A}\right)\left({}^{J}/_{m^2}\right)$
1	200	30.50	41.358	822.72
2	400	32.30	43.800	871.30
3	600	41.00	55.600	1106.03
		AVERAGE IMPACT STRENGTH		**933.35**

Table 3: Impact value for annealed specimen quenched in Water

S/N	TEMPERATURE ($°C$)	IMPACT ENERGY ABSORBED (E) Ft - Ib_f	Joules	IMPACT STRENGTH $=\left(\frac{E}{A}\right)\left({}^{J}/_{m^2}\right)$
1	200	31.80	43.121	857.79
2	400	32.50	44.070	876.67
3	600	42.10	57.088	1135.63
		AVERAGE IMPACT STRENGTH		**956.70**

Table 4: Impact value for annealed specimen quenched in diesel Oil

S/N	TEMPERATURE ($°C$)	IMPACT ENERGY ABSORBED (E) Ft - Ib_f	Joule	IMPACT STRENGTH $=\left(\frac{E}{A}\right)\left({}^{J}/_{m^2}\right)$
1	200	38.3	51.935	1033.12
2	400	42.0	56.952	1132.92
3	600	45.6	61.834	1230.04
		AVERAGE IMPACT STRENGTH		**1132.03**

Table 5: Impact value for annealed specimen normalized in air

S/N	TEMPERATURE ($°C$)	IMPACT ENERGY ABSORBED (E) Ft - Ib_f	Joules	IMPACT STRENGTH $=\left(\frac{E}{A}\right)\left({}^{J}/_{m^2}\right)$
1	200	44.3	60.071	1194.97
2	400	45.1	61.156	1216.55
3	600	46.8	63.461	1262.40
		AVERAGE IMPACT STRENGTH		**1224.64**

Graphical representation of the impact test (that is impact energy versus temperature) for the different quenching medium are as shown in Figure 5

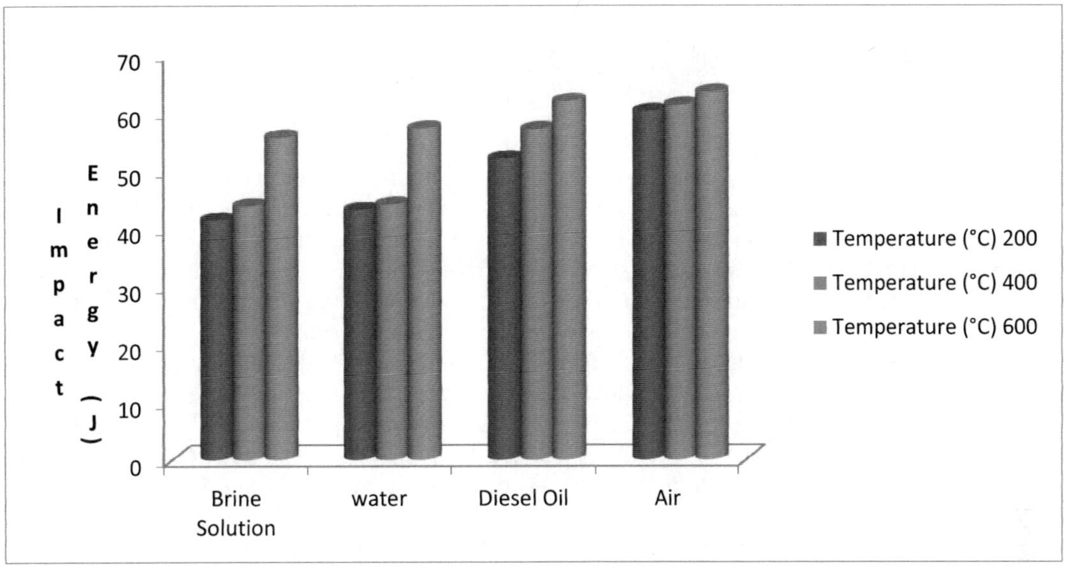

Figure 5a: Graph of Izod Impact energy versus Annealed Specimen temperature of different quenching Media

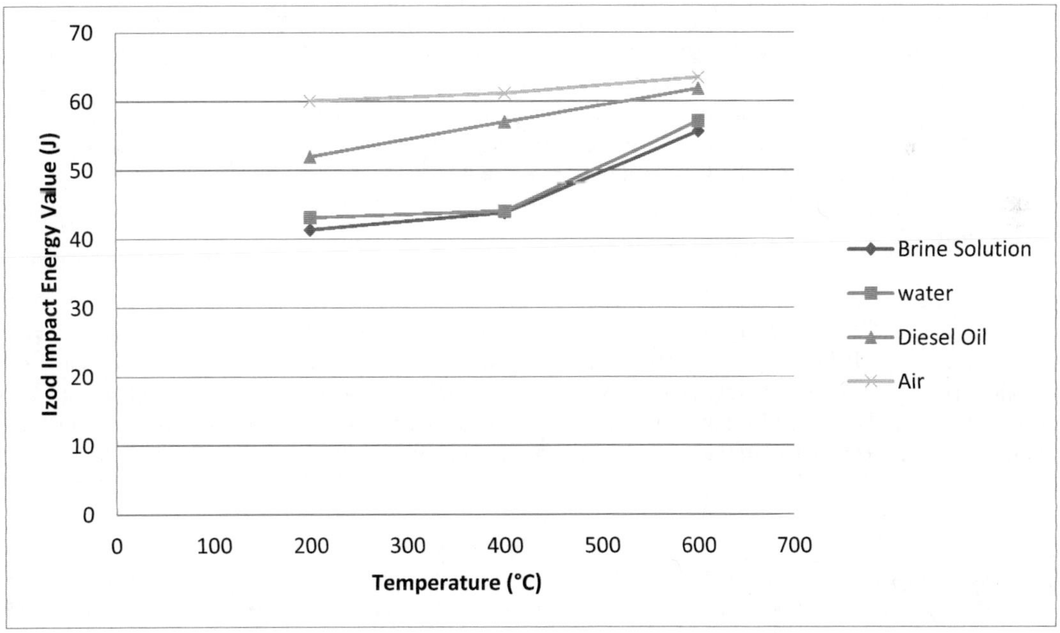

Figure 5b: Graph of Izod Impact energy versus Annealed Specimen temperature of different quenching Media

IOP Conf. Series: Materials Science and Engineering **391** (2018) 012007 doi:10.1088/1757-899X/391/1/012007

3.2 Discussion of Results

From Table 2 to 5 showing the effects of quenching on impact strength, it was observed that the impact strength increases as the temperature increases in each quenching medium; this was in line with the work of [15]. The average impact strength was higher in Air medium than diesel oil medium, diesel oil medium is higher than water medium and lastly water medium is higher than brine solution. Results show that welded mild steel normalized in air produces greatest impact strength followed by the one quenched in diesel oil and from the result the one quenched in brine solution produces less impact strength. Also impact strength of specimens quenched in diesel oil is greater than that quenched in water this was in agreement with the report made by [5, 9]; this is because the cooling rate of diesel oil is slower than water. The net result shows that the vapour transport stage is shortened, thus showing the cooling rate which of course, reduces the risk of distortion and cracking in diesel oil than water quenched.

This was also evidence in the graphical representation of the comparison of different quenching medium as shown in figure 5, Air cooled specimens at all annealed temperature had highest impact energy absorbed and were more ductile than other specimens quenched in other media, this was also supported by the work. It was observed that brine quenched specimens had the least impact energy absorbed at all annealed temperatures leading to least impact strength in all and consequently had highest brittleness.

In the same figure 5, it was observed that quenching in brine solution was brittle at lower temperature than all other quenching media. It was revealed that normalized in air had the highest ductility at all temperature range followed by diesel oil quenchant.

Figure 5 shows clearly that at annealed temperature, 200^{0}C, air cooled had higher impact energy absorbed followed by oil cooled, followed by water cooled and lastly brine solution. At annealed temperature 400^{0}C and 600^{0}C, the same trend applicable.

It was further revealed that at 400^{0}C, quenching in water and brine solution had equivalent impact energy absorbed but that of water quenched was a little bit higher impact strength than that quenched in brine solution. The Implication of this was that air cooled had the highest impact strength and more ductile than any other cooling medium and can be less prone to distortion and crack. Whereas water cooled and brine solution cooled were relatives, they had almost equivalent impact energy absorbed at each annealed temperature while oil cooled had better cooling after air cooled.

4. Conclusion

From the results obtained and their discussions, the following conclusions were drawn from the research carried out on the effects of quenching on impact strength of mild steel welded joints. Impact strength increases as the annealed temperature increases. Impact strength is higher in air cooled welded joints than oil cooled, while oil cooled was higher than water cooled and lastly water cooled was higher than brine solution cooled. That air is the best and the most economic quenching medium for welded components or parts made of mild steel subjected to impact or sudden or shocking loading.

References

[1] Groove, M.P. , Fundamentals of Modern Manufacturing: Materials, Processes and Systems. 4Th Edition, John wiley & Sons, Inc., 111 River Street, Hoboken, NJ07030 – 5774, USA, pp. 693 (2010)

[2] Parmar, R.S ., Welding Engineering and Technology. Published by Romesh Chander Khanna for Khanna Publishers, 4th reprint Edition. 2-B Nath Market, Nai Sarak, Delhi-110006, India, pp. 1, (2013).

[3] Talabi, S.I., Owolabi, O.B., Adebisi, J.A.,Yahaya, T. , "Effect of welding variables on mechanical properties of low carbon steel welded joint". Advances in Production Engineering & Management. Volume 9 I Number 4 I pp 181–186, (2014).

[4] Rassizadehghani, J.; Raygan, Sh.; and Askari, M. , "Comparison of the quenching capacities of hot salt and oil baths". Metal Sci. Heat Treat. 48(5-6): 193-8, (2006).

[5] Momoh I.M., Akinribide O.J., Ayanleke J., Olowonubi J., Olorunfemi G.O., Oshodin T. , "Investigating the Mechanical Properties of Post Weld Heat Treated *0.33%C* Low Alloy Steel". International Journal of Science and Technology Volume 2 No. 6, pp 1-5, (2013).

[6] Momoh I.M., Olateju O.O., Oloruntoba D.T. , "Investigation on the Corrosion Performance of Nickel Electrodeposited Tempered Steel Substrate". Journal of Science and Technology. Vol. 1, No.8. Pp 400 – 404, (2012).

[7] Momoh ,I.M., " Microstructures, Mechanical Properties and Corrosion Sensitivity of Dual Phase Medium Carbon Low Alloy Steel". M.Eng Thesis. Federal University of Technology, Akure, Nigeria. Pp 1 – 87, (2012).

[8] Motagi B.S and Ramesh B. , "Effect of Heat Treatment on Microstructure and Mechanical Properties of Medium Carbon Steel". International Journal of Engineering Research and Development. Vol. 2, Issue 1, Pp. 07-13, (2012).

[9] Jamiu K.O., Tajudeen K.A., Abdullarim B.R., "Evaluation of Mechanical Properties of Medium Carbon Steel Quenched in Water and Oil". AU J.T. 15(4): Pp. 218-224, (2012).

[10] Senthilkumar, T and Ajiboye, T. K., "Effect of Heat Treatment Processes on the Mechanical Properties of Medium Carbon Steel". *Journal of Minerals & Materials Characterization & Engineering*, Vol. 11, No.2 pp.143-152,(2012).

[11] Susheel ,K. S and Syed , H, M., "Influences of the welding process parameters on the welding of material" , International Journal of Engineering & Advanced Technology. Vol -2,Issue -5 , pp 22 – 29, (2013).

[12] Sameer, S . K., "Study of influence of Welding Parameters on Mild Steel". *International Advanced Research Journal in Science, Engineering and Technology.* Vol. 2, Issue 5, pp 1-3, (2015).

[13] Sunar . M; Yilbas, B.S., Boran , K., "Thermal and stress analysis of a sheet metal in welding". Journal of Materials Processing Technology vol. 172, pp123–129, (2006).

[14] Singh, B.K. , Jha, A.K. , Singh , P. K., "Effects of Joint Geometries On Welding of Mild Steel By Shielded Metal Arc Welding (SMAW)". International Research Journal of Engineering and Technology. Volume: 02, Issue: 07, pp. 1 - 7, (2015).

[15] Bodude, M. A. and Momohjimoh, I ., "Studies on Effects of Welding Parameters on the Mechanical Properties of Welded Low-Carbon Steel". Journal of Minerals and Materials Characterization and Engineering, Vol 3, pp. 142-153, (2015).

Product yield distribution and Essential Oil Composition of Eucalyptus Terminalis Sawdust Pyrolysis

J. A. Oyebanji[1*], S. O. Oyedepo[1], P. O Okekunle[2], O. A. lasode[3] and T. Adekeye[4]

[1*]Department of Mechanical Engineering, College of Engineering, Bells University of Technology, Ota, Nigeria
[1,4]Department of Mechanical Engineering, Covenant University, Ota, Ogun state, Nigeria
[2]Department of Mechanical Engineering, Ladoke Akintola University of Technology, Ogbomoso ,Nigeria
[3]Department of Mechanical Engineering, Faculty of Engineering and Technology, University of Ilorin, Ilorin, Nigeria *Corresponding Author: banjijoe@yahoo.com

Abstract. In this study, pyrolysis of sawdust sample (Western blood-wood- Eucalyptus terminalis) was investigated. Experiments were performed at six temperature levels ranging from 300 °C to 800 °C under N_2 atmosphere. The weights of char, tar and gas yields produced in each experiment were measured and recorded in percentage of initial weight of the pyrolyzed sample. Results of the study showed that product yield of Eucalyptus terminalis char, tar and gas of 41.28% at 300 °C, 45.10% at 300 °C and 57.20% at 800 °C, respectively were produced. Proximate analysis shows that volatile matter, fixed carbon, ash content and moisture content of sawdust sample were 75.53%, 15.35%, 1.56% and 7.56%, respectively. Result of the elemental analysis shows that the carbon, hydrogen, nitrogen, oxygen and sulphur contents of the sawdust sample were 54.19%, 7.05%, 0.97%, 37.15%, and 0.64%, respectively. The higher heating value and pH of the sawdust sample are 23.40 kJ/g and 2.30% respectively. This indicate that char and tar yields decrease with increased pyrolysis temperature while gas yield increases as pyrolysis temperature increases for the sawdust sample. The value of the correlation coefficient obtained indicate a fairly high degree of accuracy of the regression models to predict experimental result when used within the temperature range considered in this study. Result of analytical Py-GC/MS shows that the proportion of phenolic compounds identified was more than 50% with *trans*-2-octadecadecen-1-ol, *cis*-10-pentadecen-1-ol, 9-octadecenal and methyl-1-cyclohexenyl ketone dominating. This study establishes the fact that pyro-oil can not only be used as a fuel but can also be purified and serves as raw materials for chemical and processing industries.
Key Words: Sawdust sample, pyrolysis, pyro-oil, proximate analysis, GC-MS analysis

1. Introduction

Research into renewable energy sources are being pursued vigorously at both national and international levels. There are many factors leading to this turn of events. One compelling factor is the depleting tendency of fossil fuels, coal and natural. As a non-renewable and finite resource, fossil fuels as sources of energy are limited in supply because they are time-bound [1]. There are indications that fossil fuel will begin to dwindle out in the 21st century. Since they take millions of years to be produced, replacement of fossil fuels is a

Content from this work may be used under the terms of the Creative Commons Attribution 3.0 licence. Any further distribution of this work must maintain attribution to the author(s) and the title of the work, journal citation and DOI.
Published under licence by IOP Publishing Ltd

gradual process that cannot be relied on to continue serving future energy needs. There is concern over the security of fossil fuel supplies and their rising cost [2].

Another important issue driving the use of renewable energy is the growing concern about the environment specifically in the area of climate change. This seems to be one of the most important factors [3]. Burning of fossil fuels to produce energy leads to the release of gases, including carbon dioxide, carbon monoxide, sulphur dioxide, nitrogen oxides, methane, and other gases termed as greenhouse gases [1]. Their release into the current atmosphere causes an imbalance in the composition of these gases especially carbon dioxide and carbon monoxide leading to trapping of heat from the earth surface and a consequent increase in the atmospheric temperature of the earth termed as global warming. Renewable energy seems to be the answer to future energy needs. All sources of energy that when properly managed can be an unlimited source in terms of time are referred to as renewable energy [4]. So far, solar energy, wave energy, tidal energy, geothermal energy, hydropower and biomass energy are some of the renewable energy sources currently in use.

Thermochemical conversion of biomass is a promising route for energy and fuel production (Huber, el at, 2006), which is also considered as the easiest process to adapt to current energy infrastructure [5]. More attentions have been focused on fast pyrolysis for bio-oil production as the substituent for transported fuels [6]. The feedstock ranges from barkwood, agricultural wastes/residues, nuts, seeds, algae,grasses, and forestry residues to energy crops such as miscanthus and sorghum. Wood materials are widely used as they can give high liquid yields, up to 80% (by mass) on dry feed basis at 500-520 °C with residence time not more than 1 s.

The problem of solid waste disposal has become an alarming environmental problem facing many cities in Nigeria with lack of efficient waste management systems. However, sustainable energy can be generated from these wastes thereby putting an end to environmental pollution generated from them. In this study, the potential of bio-oil production from wood residues using pyrolysis had been examined. Pyrolysis had been considered because of its superiority to other systems of waste-to-energy conversion as it operates at atmospheric pressure and modest temperatures with yields of bio-oil exceeding 70 wt % [7].

According to [8], the price of this potential bio-oil at ₦50/litre in Nigeria is quite lower than the current price of diesel (₦180/litre) in which both are being used in many static applications such as boilers, furnaces, engines and turbines for electricity generation. Hence, adequate consideration should be given to this alternate means of energy generation thereby, reducing over-dependence on fossil fuels which are very expensive for industrial purposes. Also, Nigerian energy policy should be reviewed so as to encourage both local and foreign investors in bio oil production.

Comparatively, biomass in Africa has not received significant research attention and thus remains largely unexplored [6]. There is, however, a gradual shift occurring in favour of advanced biofuels on the African continent, both in biological and thermal conversion processes, underlined by recognition of the potential of lignocellulosic resources for the sustainable production of biofuels and biochemicals in the near future. According to [9] about 13 million tonnes of agro-forestry waste is generated annually in Nigeria.

However, the utilization of these natural resources is limited. Furthermore, with regard to fast growing population of the country and substantial waste production, waste management policy in different sectors especially biomass wastes demand novel solutions. One of the best options to reduce waste and produce green energy is energy recovery from bio-based wastes.

Presently, in Nigeria Eucalyptus terminalis is commonly used for furniture making and as building materials, it forms large percentage of wastes and constituting nuisance to the environment. Researchers in Africa are

making efforts to study the pyrolysis process for bio-oil production from different feedstock. [10] have converted corn cob, wood residue and palm fruit bunches to bio-oil, respectively.
Better management of the wastes from wood industries utilizing Eucalyptus terminalis could be achieved by producing clean energy from the materials. The clean energy is derived from pyrolysis liquid serving as source of chemical energy (fuel) and useful industrial chemicals. This study was focused on characterizing chemical composition of bio-oil yields using analytical Py-GC/MS and investigate the effects of pyrolysis temperature on product yield distribution from pyrolysis of wood residue within the temperature range of 400 °C and 800 °C in a fixed bed reactor. Regression equations were also developed from product yield curves of wood residue for the purpose of predicting pyrolytic yields.

2. Materials and Methods

2.1 Feed Materials
The wood specimen used in this study were obtained in July 2016 from Ota, South-Western Nigeria, (6° 38′ 0′′ N, 3° 6′ 0′′ E). It was air dried and ground to particle size of 500 µm using Philip milling machine.

2.1.1 Experimental Set-up
The pyrolysis unit comprised a fixed-bed reactor, retort, pipes, condenser, and a carrier gas cylinder. Fixed-bed reactor heats up the pre-loaded retort by a 4-kW furnace with automatic temperature control, hence the giving off of the volatile stream. The product pipes channel the volatile stream into the condenser which are immersed in an ice-bag (tar trapper). Staged tar trapping was employed so as to collect tarry components which escaped the first tar trap in order to ensure efficient trapping of the tar. Figure 1 shows the assembly view of the pyrolysis unit for this study.

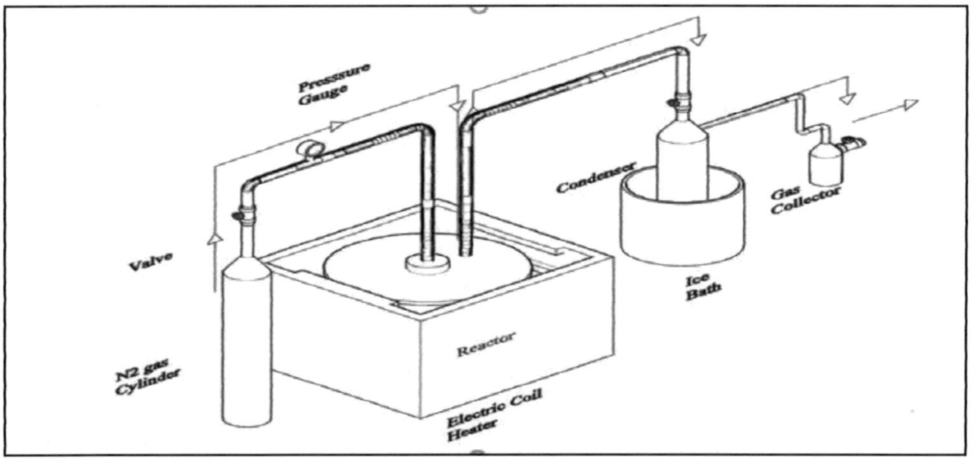

Figure1. Experimental set-up of the pyrolysis unit.

2.1.2 Experimental procedure

For the experimental investigation, six (6) temperature levels were considered i.e. 300°C to 800°C at interval of 100°C were considered in this study. Nitrogen was introduced into the reactor at pressure lower than 10 psi at a fixed holding time of ten (10) minutes and vacuum pressure. The reactor was raised to the desired pyrolysis temperature and the retort, already loaded with the sample of approximately 100 g of wood sawdust sample, was then put in the reactor with its lid firmly secured in place by using the hold down bolts with the gasket in position. The product collected were weighed to ascertain their initial weights and afterwards immersed in ice-bags. The stopwatch was set to a thirty minutes countdown. At the lapse of the thirty minutes residence time, the retort was removed from within the furnace chamber. The collected tar was measured on the weighing balance while the retort was allowed to cool. The bolts holding the retort lid in position were then loosened and a tong was used to collect the char from the retort. The char was then weighed thereafter and its value recorded. From the measured weight of tar and char, the weight of gas let off was obtained as well as the percentage weight of all products. The furnace temperature was raised again to the desired temperature and the entire process was repeated for other runs and temperature levels.

2.2 Product Analysis

The pyrolysis unit was built in such a way to assist separating the tar and gas in the volatile stream. The char remained in the retort while the fluids were transferred to condenser and gas moved to gas collector based on the simple principle of condensation. The parameters for comparison include weight of char, tar and gas produced in each experiment. A digital weighing scale of accuracy ±0.1 g was employed in measuring the weight of the samples and the products. Product yields were expressed in percentage of the initial weight of the pyrolyzed wood samples.

2.3 Compositional Analysis

Compositional analysis was conducted using both observation and instrumental analytical methods to obtain the extractive and lignin contents according to ASTM D1108-96 and ASTM D1106-96, respectively.

2.4 Heating values

The Higher Heating Value (HHV) of biomass product was calculated from a correlation developed by [11] as shown by the following equation:

$$HHV_{dry}\ (MJ/kg) = -1.3675 + 0.3137C + 0.7009H + 0.0318O^{*} \quad (1)$$

Where C, H are percentages on dry basis of carbon, hydrogen, respectively

The lower heating value (LHV) was calculated from HHV and the hydrogen content by the following equation (ECN-Phyllis, 2012):

$$LHV_{dry}\ (MJ/kg) = HHV_{dry} - [2.442 \times 8.936(H/100)] \quad (2)$$

3. Results and Discussion

3.1. Physical characterization

The results were discussed in terms of the physical and chemical properties of the solid raw sample. The main characteristics of the feed material are listed in Table 1.

Table 1. Main characteristics of the raw wood sawdust sample (%)

Proximate analysis[a] (wt. %)		**Ultimate analysis[b] (wt. %)**	
Moisture content	7.56	Carbon	54.19
Volatile matter	75.53	Hydrogen	7.05
Fixed carbon	15.35	Nitrogen	0.97
Ash	1.56	Oxygen[c]	37.15
pH	2.30	Sulphur	0.64
HHV(MJ kg^{-1})	23.40		
LHV(MJ kg^{-1})	17.35		

3.2 Product yields

The product yield distribution from the pyrolysis of Western blood-wood (Eucalyptus terminalis) sample at different temperature levels are presented in Table 2, respectively. As shown in the Table, char and tar yields from the wood sawdust pyrolysis decreased from 41.28% and 45.10% at 300 ^{0}C to 24.15% and 18.65% at 800 ^{0}C, respectively. However, gas yield increased from 13.62% at 300 ^{0}C to 57.20%.

Table 2: Product yield distribution from Western blood-wood (Eucalyptus terminalis)

Temperature (^{o}C)	**Char yield (wt %)**	**Tar yield (wt %)**	**Gas yield (wt %)**
300	41.28	45.10	13.62
400	32.52	37.05	30.43
500	28.40	34.25	37.35
600	25.80	30.45	43.75
700	25.45	25.30	49.25
800	24.15	18.65	57.20

3.3 Effect of Temperature on the Product Yields

Figure 2 shows product (tar, pyro-gas and char) yield at different pyrolysis temperature. From Figure 2, it can be seen that tar and char yield decreases with increase in temperature. This is because bio-oil secondary reactions yields more gas and char at temperatures above 500 ^{0}C. This is similar to results of Fagbemi et al. (2001).Char yield decreases from 41.28 wt% to 24.15 wt% with increased pyrolysis temperature. This may be due to increase rate of primary decomposition at high temperature or through secondary decomposition of the solid residue. This result is consistent with that of [12]. and Gheorghe [13]. It is observed that as temperature increases from 300^{0}C to 800^{0}C the bio-gas yield increases from 13.62 wt% to 57.20 wt%.

IOP Conf. Series: Materials Science and Engineering **391** (2018) 012008 doi:10.1088/1757-899X/391/1/012008

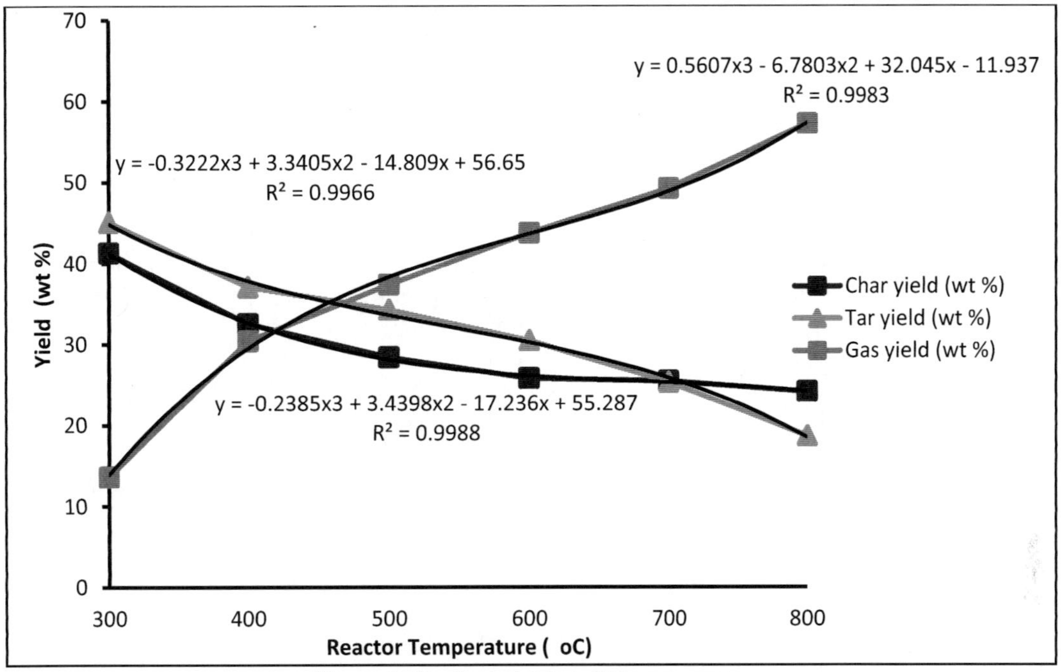

Figure 2. Product yield at different reactor

3.4 Regression Models between Reactor Temperature and Product Yields

The regression models between the reactor temperatures and product yields, and the respective value of the coefficient of correlation (R^2) for Western blood-wood (Eucalyptus terminalis) were obtained and are given as

Char: $Y_c = -0.2385T^3 + 3.4398T^2 - 17.236T + 55.287$ $R^2 = 0.9988$ (3)

Tar: $Y_t = -0.3222T^3 + 3.3405T^2 - 14.809T + 56.65$ $R^2 = 0.9966$ (4)

Gas: $Y_g = 0.5607T^3 - 6.7803T^2 + 32.045T - 11.937$ $R^2 = 0.9983$ (5)

Where Y_c is the char yield, Y_t is the tar yield, Y_g is the gas yield and T is the reactor temperature. The models (equations (3) to (5)) are instrumental for predicting the response of wood feedstock pyrolyzed in a fixed-bed reactor between 300°C-800°C. The values of the correlation coefficients obtained indicate a fairly high degree of accuracy of the regression models to predict experimental results when used within the temperature range considered.

3.5 Composition of the Volatile oil of Western Blood-wood (Eucalyptus terminalis).

The bio-oil yields obtained from the pyrolysis of the wood sample were subjected to Gas Chromatography (GC) and Gas Chromatography-Mass Spectrometry (GC-MS) analyses. This was done for detailed identification of various compositions in the complex mixture. These compositions are presented in Table 5. The retention indices of each identified component were also calculated based on their retention time in order to confirm the identification.

The GC-MS analysis of the bio-oil yield revealed the presence of of 21 compounds (99.4% of the volatile extract). The main components are palmitic acid (13.0%), oleic acid (16.3%), cis-1, 9-hexadecadiene (12.5%), *cis*-10-pentadecen-1-ol (12.4%), 9-octadecenal (10.9%), *trans*-2-octadecadecen-1-ol (10.0%), myristic acid (5.1%), stearic acid (5.0%) and other compounds (14.8%).

Table 3: Chemical composition

No.	Compounds	Retention Index	Percentage Composition	Group
1	Luprosil	676	4.0	$C_3H_6O_2$
2	2,5-dimethylfuran	732	2.3	C_6H_8O
3	1,4-dimethyl-1H-pyrazole	804	0.7	$C_5H_8N_2$
4	2-methylbutyric acid	811	1.2	$C_5H_{10}O_2$
5	*trans*-3-Octene	824	0.3	C_8H_{16}
6	*cis*-3-octene	824	0.3	C_8H_{16}
7	5-methyl-3-methylene-5-hexen-2-one	887	0.6	$C_8H_{12}O$
8	4,5-dimethyl-1H-imidazol	927	0.6	$C_5H_8N_2$
9	Cyclooctane	959	0.3	C_8H_{16}
10	Corylone	972	0.5	$C_6H_8O_2$
11	3-methylcyclopentane-1,2-dione	1003	0.5	$C_6H_8O_2$
12	methyl-1-cyclohexenyl ketone	1027	1.2	$C_8H_{12}O$
13	*o*-guaiacol	1090	0.3	$C_7H_8O_2$
14	2-nitroethylpropionate	1067	0.8	$C_5H_9\,NO_4$
15	*cis*-1,9-hexadecadiene	1610	12.5	$C_{16}H_{30}$
16	*cis*-10-pentadecen-1-ol	1763	12.4	$C_{15}H_{30}O$
17	myristic acid	1769	5.1	$C_{14}H_{28}O_2$
18	palmitic acid	1968	13.0	$C_{16}H_{32}O_2$
19	9-octadecenal	2007	10.9	$C_{18}H_{34}O$
20	*trans*-2-octadecadecen-1-ol	2061	10.0	$C_{18}H_{36}O$
21	stearic acid	2167	5.0	$C_{18}H_{36}O_2$
22	oleic acid	2175	16.9	$C_{18}H_{36}O_2$
	Percentage Total		**99.4**	

4. Conclusion

In this study, the potentials of waste-to-energy system had been assessed by considering the production of bio-oils from wood residue collected from south-western Nigeria using pyrolysis conversion. The chemical compositions of fast pyrolysis bio-oils from wood have been studied by GC-MS analysis and a fairly high

degree of accuracy of the regression models to predict experimental results was obtained. Results of this show that the bio-oil produced met the qualities of other liquid products from other energy classified woody biomass already screened. The difference in level of phenolic compounds, oleic acid, aromatic hydrocarbons and nitrogen-contained compounds suggest that it could be a good source of natural antioxidants to be used in medicinal and food products to promote human health and prevent diseases,
This study establishes the fact that understanding the compositions of bio-oils from energy crop will add to the use in thermochemical conversion platforms, as this information will be extremely important in evaluating the bio-oils' stabilities, properties, and toxicities. It will also be useful in identifying and solving challenges associated with further refining of bio-oils into liquid fuels and that selected sawdust sample have the potential of being used as alternative source of energy and serve as good source of extraction of chemical compounds.

Acknowledgements
The authors acknowledge the technical support of Ololade Z.S with the GCMS interpretations.

References

[1] Adegoke, O.A., Fuwape, J. A., Fabiyi, J. S. (2014), Combustion properties of some tropical wood species and their pyrolytic products characterization. *Energy and Power*. 4(3):54-57.
[2] American Standard of Testing and Materials, 1996. ASTM D1106-96, 2007. Standard Test Method for Acid-insoluble Lignin in Wood. ASTM International, West Conshohocken, PA.
[3] American Standard of Testing and Materials, 1996. ASTM D1108-96, 2007. Standard Test
[4] Balogun, B.O and Salami, A.T (2016), 'Effects of biofuel production on selected local Communities in Nigeria', *Journal of Petroleum Technology and Alternative Fuels,* Vol. 7(3), pp. 18 – 30
[5] Bensah, E.C., Kadar, Z., and Mensah, M.Y. (2015). Ethanol production from hydrothermally-treated biomass from West Africa. *Bio Resources* 10(4): 6522-6537.
[6] Budarin, V.L., Clark, J.H., Lanigan,B.A., Shuttleworth, P., Breeden,S.W., Wilson, A.J., Macquarrie, D.J., Milkowski, K., Jones, J.,Bridgeman, T., Ross, A.,(2009). "The preparation of high-grade bio-oils through the controlled, low temperature microwave activation of wheat straw", *Bioresour. Technol.,* 100, 6064-6068
[7] Fagbemi L., Khezami l., and Capart. L. (2001). Pyrolysis products from different biomasses: application to the thermal cracking of tar. *Applied Energy* 69, 293–306.
[8] Gheorghe C. B., Mărculescu. C., Badea. A., and Apostol,T. (2010). Pyrolysis Parameters Influencing the Bio-Char Generation from Wooden Biomass. *U.P.B. Science Bulletin. Series* C, 72, 1
[9] Harvey, P. A. (2004). Chem E interfaces-The Research Newsletter of the Dept. of Chemical Engineering. University of Cambridge. Issue 6, Jan.2004.
[10] Haq,Z.(2002).Biomass for Electricity Generation,Energy Information Administration,feature article available at http://www.eia.doe.gov/oiaf/renewable.html.
[11] Huber, G.W., Iborra, S., Corma, A. (2006). "Synthesis of transportation fuels from biomass: Chemistry, catalysts and engineering", *Chem. Rev.,* 106, 4044-4098.
[12] Kuye, A.O., Edeh. I. (2013). Production of bio-oil from biomass using fast pyrolysis – A critical review. *Journal of Minerals Research*. 1(1): 1-20.

[13] Lédé, J., Broust, F., Ndiaye, F.T., Ferrer, M., (2007). "Properties of bio-oils produced by biomass fast pyrolysis in a cyclone reactor", Fuel, 86, 1800-1810.

[14] Ogunjobi, J.K., and Lajide, L. (2013). Characterisation of bio-oil and bio-char from slow pyrolysed Nigerian yellow and white corn cobs. *Journal of Sustainable Energy and Environment*. 4:77-84.

[15] Ogunsina, B.S., Ojolo, S.J., Ohunakin, O.S., Oyedeji, O.A., Matanmni, K.A., and Bamgboye, I.A. (2014). Potentials for generating alternative fuels from empty palm bunches by pyrolysis. JOMAR.8:50-60.

[16] Olah, G.A., Alain, G. and Prakash, S.G.K, (2006). Beyond Oil and Gas: The Methanol Economy. Wiley-VCH Weinheim 2006.

[17] Picchi, G., Gordon, A. and Naresh, T., (2007). Feedstock to Furnace: Bioenergy Systems for the Ontario Greenhouse Industry. Potential in Ontario for the Implementation of Willow Short Rotation Coppice Cultivation for the Greenhouse Industry.

[18] Popoola, L.T, Gutti, B, Adeniran, J.A and Adeoye, B.K (2013), 'The potentials of waste-to-energy system in Nigeria: A study of pyrolysis conversion of wood residue to bio-oil in major cities of south-western Nigeria', *Advances in Applied Science Research*, 4(2):243 – 251

[19] Sheng, C.D., Azevedo, J.L.T. (2005). Estimating the higher heating value of biomass fuels from basic analysis data. *Biomass & Bioenergy*. 28,499-507.

[20] Wagernaar, B.M., Prins, W. and Van Swaaj. W.P.M. (1993). Flash Pyrolysis Kinetics of Pine Wood. *Fuel Processing Technology*, 36:291-298

Assessment of Some Clay South West and Aluminmium Dross as Roofing Tile Materials.

A. O. Inegbenebor[1*], O. S. I. Fayomi[1], O. Joshua[2], K. J. Jolayemi[3], A. I. Inegbenebor[4], O. O. Joseph[1]

[1*]Department of Mechanical Engineering, Covenant University, P.M.B 1023, Ota, Nigeria. Corresponding Author.
[1]Department of Mechanical Engineering, Covenant University, P.M.B 1023, Ota, Nigeria.
[2]Department of Building Technology, Covenant University, P.M.B 1023, Ota, Nigeria.
[3]Department of Civil Engineering, Covenant University, P.M.B 1023, Ota, Nigeria.
[4]Department of Chemistry, Covenant University, P.M.B 1023, Ota, Nigeria.

Abstract- Clay has been widely used as a raw material in construction and in the building sector from the ancient period till date. It is a sustainable material and the technology of processing it can be found anywhere in the world. As a result of its sustainability, it can be used in many industries. One of the products derived from clay is roofing tiles. Some of the properties to be considered are thermal and durability of the materials, are necessarily to be taken into consideration, especially if the building needs to be situated in tropical region. Clay seems to be a good choice. This paper will therefore, assess the possibility of using clay from south west and aluminum dross as composite roofing tiles. Clay samples were collected from five (5) deposits in Igboora (Oyo), Ilaro, Itori, Ajegunle in (Ogun) and Ikorodu in (Lagos) states in south west of Nigeria. The clay samples were crushed, pulverized and sieved. Their chemical compositions analysis was carried out using Atomic Absorption Spectrometer (AAS); molds 135mm by 120mm by 15mm were made for the clay samples. Composites were made in the following proportions by a mix of clay samples with aluminum dross in a percentage ratio of 50:50 and another mixture of clay samples with aluminum dross and bentonite as a binder at a percentage ratio of 45:45:10. The physical properties such as water absorption and cold crushing strength were carried out. Also thermal conductivity was carried out on the sample. The result of chemical analysis showed that Igboora, Silica (47.9%) an alumina (36.7%), the rest were the traces or the other element. Ilaro samples were silica (43.5%) and alumina (34.2%) Itori samples, silica (49.8%) and alumina (30.7%). Ajegunle sample, silica (46%) and alumina (38.2%) and Ikorodu sample, silica (54.5%) and alumina (24.6%). These results were compared to the Grim-Shaw specifications which specified (25.45%) alumina and 40.6% silica compositions in requirements for clay soils ascertained that the samples were all clay soils, though the Ikorodu samples was lacking a little in its alumina content. The result of water absorption test of Igboora samples and its composite agree with the Indians roofing tiles' standard (IS 654:1992) while the other samples failed. The result for Ilaro composite 50:50 had the lowest thermal conductivity of 0.086w/m.kg for the crushing strength test, the Itori and Ilaro composite with bentonite acting as binder produced results of 1.07KN and 1.01KN which agrees with the Indian roofing standard. At this preliminary work, it can be suggested that in conclusion, Igboora and Ilaro clay and their composites seem to be promising for roofing tiles.

Keywords: Roofing Tiles, Clay, Aluminum Dross, Raw Materials, Thermal Properties.

Content from this work may be used under the terms of the Creative Commons Attribution 3.0 licence. Any further distribution of this work must maintain attribution to the author(s) and the title of the work, journal citation and DOI.
Published under licence by IOP Publishing Ltd

1. Introduction

Clay is a viable mineral that exhibit high economic significance and usefulness that affects every sphere of life though it is a product of disintegration form chemical weathering of feldspathic and granite rocks [1]. It consists of fine particles of hydrous aluminum silicate which has a wide variety of physical characteristics such as plasticity, shrinkage under firing and under air-drying, fineness of grain, color after firing. Hardness cohesion and capacity of the surface to take decoration [2].
Furthermore, clay is a unique form of soil and it is of different forms. It is made up of different minerals. Clay minerals are made up of different components, among which are mineralogical, chemical composition, phase constitutions, and microstructural morphology of these materials make clay to be unique and suitable as industrial minerals whose application is dependent on its structure and chemical composition [3].
The major clay in Nigeria are three different deposits of Kaolinite, microcline, muscovite/iolite, plagioclase/ albite and quartz.

In their work [4], it was observed that the two major types of clay are the foundation of three basic types of clay bodies; earthenware, stoneware and porcelain whose maturing temperature ranges between 750°c and 1150°c, 1150°c and 1350°c as well as 1400°c and 1700°c respectively; these clays form the rudiments for other clay bodies. However, clay body formations which are varying from place to place have been documented by these authors [5]. Which also indicated the differences in the properties of the clay that are being used in industrial set up. The important properties of cay are plasticity, colour, clay strength, drging and firing shrinkages. The percentages of minerals oxides such as (Fe_2O_3, MgO, CaO, Na_2O and so on) in clay ultimately determine the areas of application of the clay such as bricks, floor tiles, paper and so on, while the quantity of alkali metal oxides (Na_2O, K_2O, CaO, etc) indicate their suitability for making ceramics products [6].
Also, one of the products from clay is roofing tiles. Roof tiles in houses have been an ancient tradition and the current use allows the maintenance of the architectural outlook in many countries. From the ancient time till present day, a wide variety of materials for roofing will afford better bargains in terms of finiancial costs, the kind of building to be erected, the availability of the material, environmental and climate factors [7]. Also thermal and durability properties of the materials are necessarily to be taken into consideration, especially if the building needs to be situated in tropical regions. Clay seems to make a good choice.
Due to its inherent properties, as a roofing material clay possesses the ability to absorb heat projected onto its surface and still leave the interior of its building structure cool. The roof receives the most solar radiation of any building component under normal circumstances [8].
Solar reflectance is the most important characteristic of a roof product in terms of yielding the highest energy savings during warmer months [9]. The higher the solar reflective value, the more efficient the product is in reflecting sunlight and heat away from the building and reducing roof temperature. This is particularly important in warm areas where peak load is a concern [10].
This paper will therefore, assess the possibility of using some clay from south west and aluminum dross as composite roofing tiles.
Aluminum dross is the by-product of the smelting process in the creation of aluminum from bauxite (an aluminum ore). It forms on the surface of the molten metal and is as a result of oxidation of the metal. Aluminum dross is a by-product which constitutes nuisance to the aluminum industries as slag to the steel industries. If the dross is found useful in roofing tile industry, it will help to keep the aluminum industries environment clean and economical.

2. Materials and Methods

a. Raw Materials

Representative samples of the five (5) deposits in Igboora in Oyo, Ilaro, Itori, Ajegunle in Ogun and Ikorodu in Lagos states, all in the south west of Nigeria. The clay samples were crused, pulverized and sieved. Their chemical compositions analysis were carried out using Atomic Absorption Spectrometer(AAS), moulds of 135mm by 120mm by15mm were made for the clay samples

Composites were made in the following proportions by a mix of clay samples with aluminum dross in percentage ratio of 50:50 and another mixture of clay samples with aluminum dross and bentonite as binder in a percentage ratio of 45:45:10. The physical properties such as water absorptions and cold crushing strength were carried out. Also thermal conductivity was carried out on the samples.

b. Cold Crushing Strength

The molded samples were fixed at 105°c for 24 hours and were tested for crushing strength using hydraulic strength machine. The crushing strength was then calculated using the relationships:

$$\text{Cold crushing strength} = \frac{load\ KN_2}{area\ (M)} \quad 1$$

c. Water Absorption Measurements

Water absorption measurements were carried out according to British standard EN 99: (1991). The fired molded samples were soaked in boiling water in a beaker and left to cool down to room temperature still soaked in water. Excess water was removed from the samples surface using a moistened cloth prior to weighing. The wet weight (W_w) was then measured. Samples were left drying in an oven at 110°C over night, and the dry weight was measured (W_d). Water absorption (W_a) was computed as:

$$Wa = \frac{Ww - Wd}{Wd} \quad 2$$

d. Thermal Conductivity Measurements

The name of the apparatus is KD2 Pro. Using KD2 pro to take measurements, appropriate sensor was chosen and attached to KD2 pro meter, and then the meter was turned on. The sensor was properly inserted into the clay molded sample. Each measurement displayed in-situ temperature of the sample, thermal conductivity and thermal resistivity respectively. After each measurement, the instrument was allowed to rest for 25 minutes before taking the next reading in order to allow equilibrium position to be established. This procedure was repeated for the remaining samples.

3. Results and Discussion

e. Compositional analysis of the clays

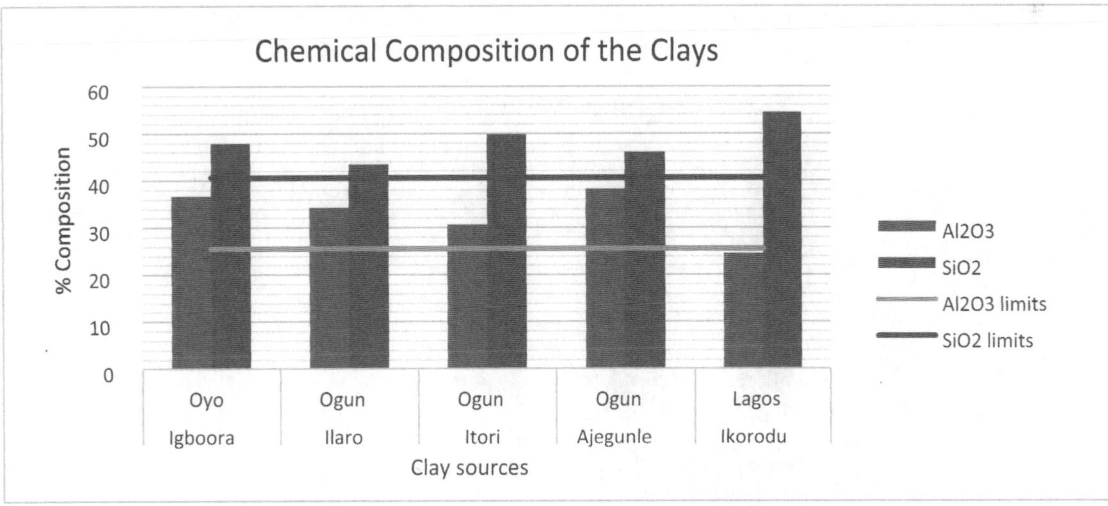

Figure 3.1: Chemical composition analysis of the clay from Southwest region of Nigeria.

f. Water Absorption Test

In the water absorption test, only the Igboora, Ajegunle and Ilaro clay samples and its composite produce the results which are in Figure 3.2.

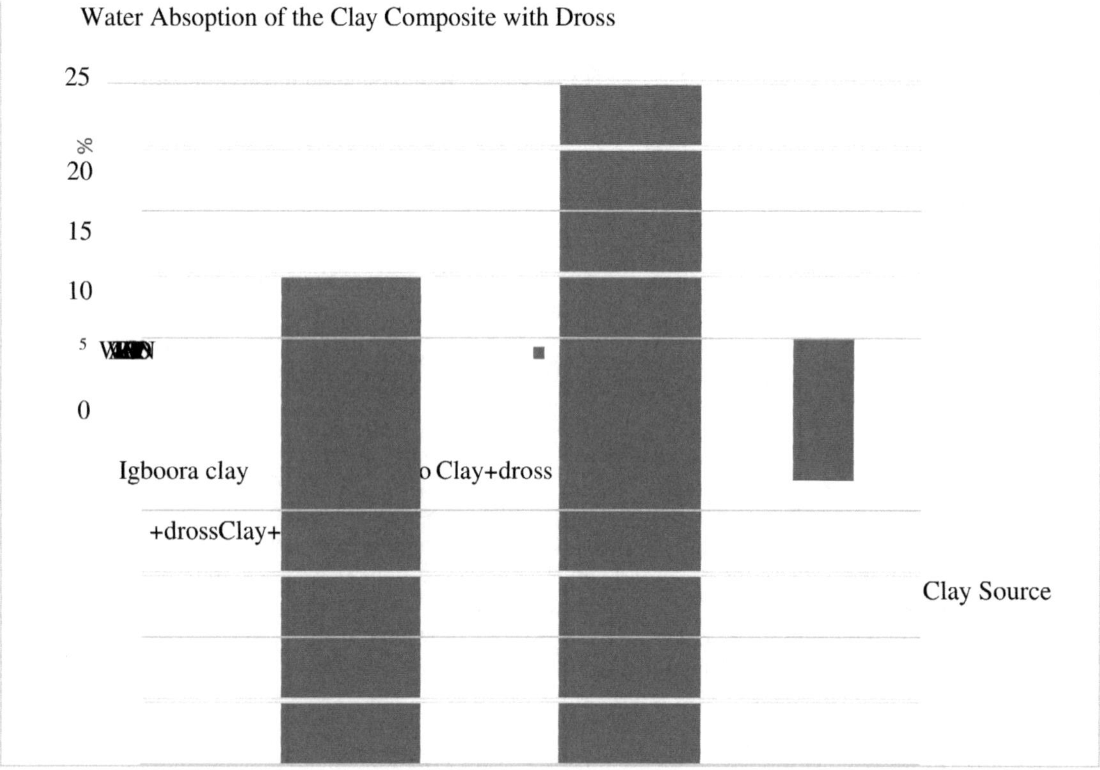

Figure 3.2: Water Absorption Results.

g. Cold Crushing Strength Result

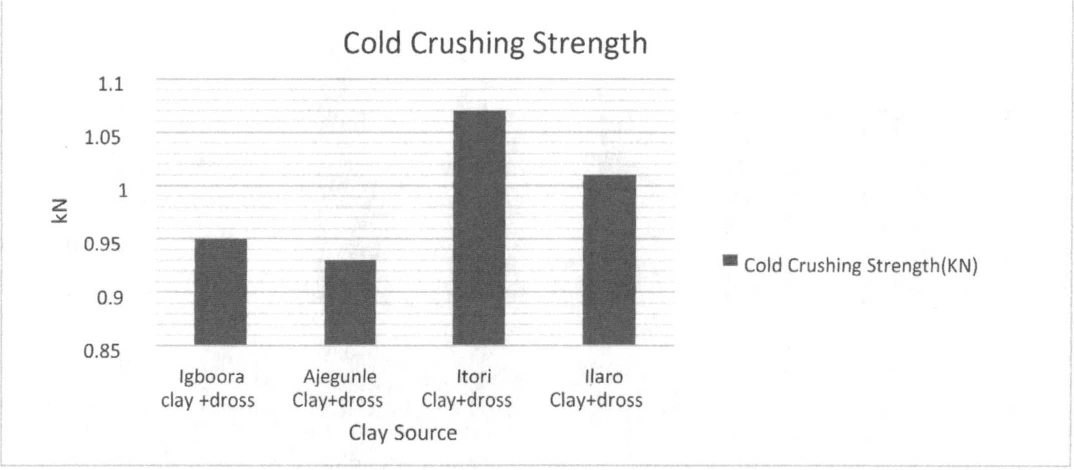

Figure 3.3 Cold Crushing Strength Result

3.4. Thermal Conductivity (W/M.K) Result.

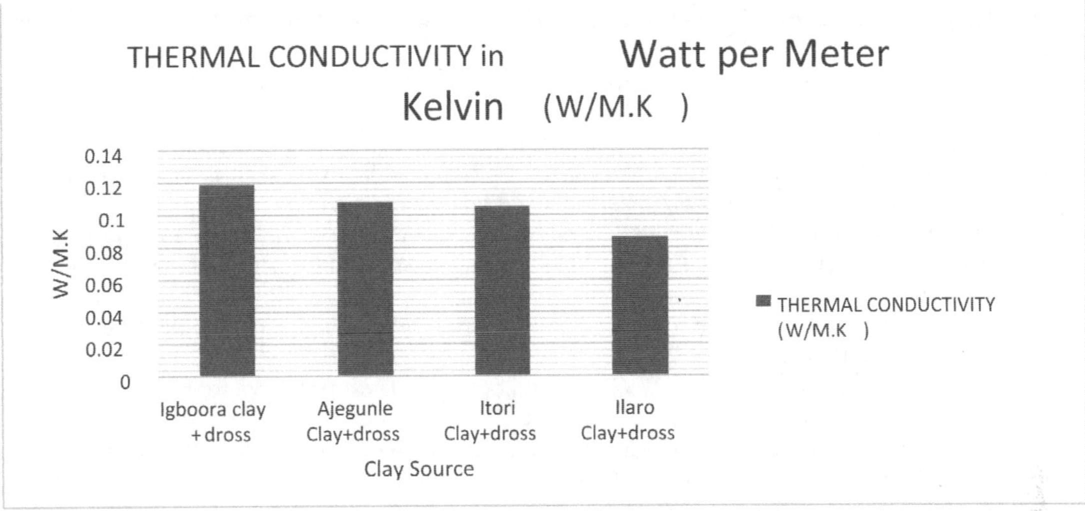

Figure 3.4 Thermal Conductivity Result.

4. Discussion and Conclusion

The results of the chemical analysis were compared to the Grim Shaw specification which specified 25.45% aluminum and 40.6% silica composition requirement for clay soils ascertained that the samples were all clays soils (although the Ikorodu samples was lacking a little in its alumina content). It failed other tests carried out on its samples. The results of water absorption test of Igboora samples and its composite agree with the Indian roofing tiles standard (IS 654:1992) [11] while the other samples failed. The Ilaro composites sample had the lowest thermal conductivity, which was unable to fit to Indian and Australia International standard [12] of roof tiles.

However, for the crushing strength test, the Itori and Ilaro composite with bentonite acting as binder produced the results which agree with the Indian roofing standard [11]. The work so far is encouraging.

5. Conclusion

At this preliminary work, it can be suggested that in conclusion, Igboora and Ilaro clay and their composites seem to be promising for roofing tiles. Further works is still ongoing to ascertain the possibility in using them on roofing tiles.

REFERENCE

[1] Adelabu, O. S., & Kashim, I. B. (2010, November). Clay mineral: A case study of its potentialities in selected parts of Kaduna State of Nigeria. In *Education and Management Technology (ICEMT), 2010 International Conference on* (pp. 655-659). IEEE.

IOP Conf. Series: Materials Science and Engineering **391** (2018) 012009 doi:10.1088/1757-899X/391/1/012009

[2] Akhirevbulu, O. E., Amadasun, C. V., Ogunbajo, M. I., & Ujuanbi, O. (2010). The Geology and Mineralogy of Clay Occurrences Around Kutigi Central Bida Basin, Nigeria. *Ethiopian Journal of Environmental Studies and Management*, *3*(3).

[3] Aramide, F. O., Alaneme, K. K., OlubambI, P. A., & Borode, J. O. (2014). Characterization of some clay deposits in South West Nigeria. *Leonardo Electronic Journal of Practices and Technologies*, *25*(2), 46-57.

[4] Kalilu, R. O. R., Akintonde, M., & Ayodele, O. (2006). Ceramics: Art and technology in the 21st century Southwestern Nigeria. *Agege: Pemilter.*

[5] INGER, F., & SINGER, S. (1963). Industrial ceramics. London: Chapman and hall LTD p649.

[6] Nnuka, E. E., & Enejor, C. (2001). Characterization Nahuta Clay for Industrial and Commercial Applications. *Nigerian Journal of Engineering and Materials*, *2*(3), 9-12.

[7] Kolawole, F. O., Adeniji, S. A., Idowu, A. T., Owoseni, T. A., Ngasoh, O. F., & Soboyejo, W. O. (2014). Corrugated Laterite Based Ceramic Roof Tile Stabilized with Cement. International Journal of Engineering and Technology 4(3), 145-149

[8] Kristin, L.G and Bradley, R (2006). The role of extensive green roofs in sustainable development. Hortscience 41(5), 1276-1285.

[9] Akbari, H, and Matthews, H.D, (2012). GLOBAL cooling updates: reflective roofs and pavements energy building, 55 pp 2-6.

[10] Akbari ,H,Berdahl,P,Levinsion,R,wiel,S,Miller,W, and Desjarlais,A,(2006) cool-color roofing material, California energy commission PIER Building End- use energy efficiency program CEC - 500-2006-067.

[11] Bureau of Indian standards (200) "clay roofing tiles", Mangalore pattern- specification (third revision). International standard of Indian (IS.654:1992).

[12] Roof tiles of association of Australia (2015) "the properties off roof tiles" pp.6.

Lifecycle Engineering of Infrastructure: An Essential Approach to Engineering for a Sustainable Africa

I.S. Dunmade[1,2] and OSI. Fayomi [2,3]

[1]Earth & Environmental Sciences Department, Mount Royal University, Calgary, Canada
[2] Mechanical Engineering Department, Covenant University, Ota, Nigeria
[3]Department of Chemical, Metallurgical and Materials Engineering, Faculty of Engineering and the Built Environment, Tshwane University of Technology, Pretoria, South Africa.

Abstract. Looking at African Development in comparison with other continents, it can be concluded that there are still so much to do to achieve comfortable standard of living for the majority of the inhabitants. While a lot of steps has been taken and appreciable progress has been made in many places, majority are yet to have access to stable and affordable basic necessities of life such as potable water, stable electricity and comfortable accommodation. This study examined various engineering approaches being used and those that could be used to make the enumerated basic necessities of life accessible to majority of Africans at affordable price. Our evaluation revealed appropriate lifecycle engineering as a feasible approach to achieving the desired goal. This research and its findings will assist Africans and other nations in understanding African infrastructure problems and how best to address the technical problems in a sustainable manner.

Keywords: Lifecycle Design, Lifecycle Management, Sustainable Africa, Sustainable Engineering, Sustainable Infrastructure, System's Lifecycle Extension

1. Introduction

The quest for improved standard of living in the African continent is enormous. The challenge is formidable and concerted effort being made to achieve the fit is strong and consistent. Progress has been made in various sectors of the economy, but the goal is yet to be achieved. A number of successes attained has been short lived by unexpected failure in some realms. The standard of living of a population is largely dependent on the strength/capacity of infrastructure asset that is in the place, effectiveness of its utilization and its maintenance [1], [2]. The infrastructural development in Africa has been heavily dependent on foreign help. The foreign help has been in terms of technology importation funded by loans and foreign aids. While foreign technology and its international diffusion is a well-recognized factor in the rapid industrialization of newly industrialized countries, it doesn't seem to yield similar result in Africa [3]. There are evidences that significant number of infrastructure technologies shipped to Africa were either based on outdated engineering technologies or technologies considered inferior to acceptable standard at the source. Many of the technologies are foreign to African culture and way of life. Consequently, they failed to achieve their intended purpose or the performances are short lived. There is therefore a need for a new approach to engineering African infrastructure in order to achieve sustainable development.

Content from this work may be used under the terms of the Creative Commons Attribution 3.0 licence. Any further distribution of this work must maintain attribution to the author(s) and the title of the work, journal citation and DOI.
Published under licence by IOP Publishing Ltd

1.1 Objective of the study

The main goal of this study is the identification of viable solutions to the long standing infrastructure engineering problem in the Sub-Saharan Africa. It involves looking for the sustainable solution that would guarantee continuous growth and prosperity of the continent Africa.

1.2 Research Questions

The research addresses the following questions:

- What are the technical issues affecting African infrastructure?
- What are the available engineering approaches for infrastructure development?
- Which engineering approach is considered appropriate for solving African infrastructure problem and why?

2. Research Methodology

The methodology of the study involved thorough examination of the causes of the longstanding infrastructure problems in Africa as identified by notable scholars, policy makers and technocrats. The study of the causes of African infrastructure problems also involved personal experiences and observations over a long period of time. This was followed by the study of various infrastructure development approaches in the developed and other developing countries, identifying those that worked well and those that did not. It also involved a study of the current progression in engineering on approaches for sustainable infrastructure development. With these knowledges of causes of African infrastructure predicament and the best practices in infrastructure engineering, an engineering approach considered suitable for African infrastructure development now and into the future was proposed.

3. Findings on African Infrastructure Problems

Our study revealed that there are quite a number of technical and socio-economic issues affecting the development, availability, smooth operation and longevity of African infrastructure [4-9]. Quoting a number of UN reports, [5] gave a vivid picture of the state of African infrastructure and how it is affecting its economic development. He stated that "Africa's poor infrastructure is slowing its economic development. African countries need to promote industrial development to spur economic progress and reduce poverty. Africa's share of global manufacturing is drastically disproportionate to its population. While 15 percent of the world's population lives in Africa, only about one percent of global manufacturing takes place there. That is largely due to poor transport, communications and energy infrastructures." The following are some of the techno-economic issues identified in the course of the study.

3.1 Restricted customer access to service

Infrastructure is a capital intensive investment. Limitation in locally generated revenue by many African governments has made many of them to depend on loans and subsidies to fund currently available infrastructure. Most of the infrastructure for energy and water supplies and transportation in many African countries are only available to urban dwellers. Majority of the rural populace does not have access to them. This problem is affecting productivity and affecting standard of living of people. This problem would have to be overcome in order for Africans to achieve the sustainable development goals on eradication of poverty, clean water and economic growth [7].

3.2 Affordability issues

Majority of the agencies providing vital services such as energy, water, telecommunication and transportation are state owned. They are usually monopolies. Many of them are so large that their operations are sub-optimal. At the advent of privatization campaign, most of them were dismembered and privatized as independent organizations with public-private participation. As a result of their arms-length operation, their service rate was jacked up but unfortunately such increase has made them become unaffordability for the ordinary citizens.

3.3 Dissatisfactory customer service

Transportation, communication, energy and water services in many countries of Africa are characterised by what can be termed "epileptic and poor quality of service". The monopolistic nature of the various service agencies in African countries made them become very large and made many of them operate like government bureaucracy. Their services are essentially unreliable as they are "off and on" over time. For instance, some municipalities have electric power supply for 2 - 4 hours in a day while some receive such service every day, a number of the cities receive such service two or three days in a week while some other ones or part of a city may not even receive electricity service for weeks. During the period of availability, many households and businesses do experience power fluctuation between low voltage and high voltage which has caused significant damages to equipments and machinery. This has also reduced productivity and consequently affected the profitability of businesses. The same epileptic and poor service is experienced by many communities in the areas of water supply and other services.

3.4 Frequent breakdowns

Unavailability of services or epileptic nature of services rendered were as a result of frequent facility/equipment breakdown. Several reasons can be adduced to causing the breakdowns. The breakdowns arose from the differences in the service conditions of the technology originating country in comparison with the country of deployment. Many of the imported infrastructure technology were from Europe and North America which are essentiality temperate climate with the attendant soil and vegetation conditions compared to the mainly tropical African climate. In addition, cultural behaviour and attitude of the technology originating country are totally different from the recipient African countries. As a result, the normal care of the system which is a given in the originating country is totally absent in the recipient country.

In addition, the frequent breakdowns arose from inadequate maintenance of the facilities either because of the replacement parts unavailability, non-availability of needed fund to procure necessary parts or negligence on the part of facility management. Significant number of the infrastructure is in a state of disrepair and breakdown. Other reason for the frequent breakdown is lack of foresight on the part of the planners and other stakeholders in ensuring the localisation of repair and maintenance capacity building into the system development strategy.

4. Some of the utilizable engineering approaches to solving African systemic infrastructure problems

Over the years, engineers have developed various design approaches to engineering systems development. Most of the modern design approaches has many things in common except few differences that attempt to eliminate the lapses of the earlier ones and to incorporate new knowledge and developments in the field [10-12]. Many of them are called different names, probably due to differences in the sources of idea and premises from where they emerge. The following are some of the modern design approaches that could be adapted to solve the longstanding African infrastructure problems:

4.1 Concurrent Engineering

According to [13], "Concurrent engineering (CE) is an engineering management philosophy and a set of operating principles that guide a product development process through an accelerated successful completion." It is also known as simultaneous engineering. It is an effective method for managing the development of complex systems; but it requires a set of analytic tools and procedures to operationalize its concepts. It involves the completion of all aspects of design process simultaneously. It promotes the incorporation of downstream concerns into the upstream phases of a development process. This would lead to shorter development times, improved product quality, and lower development– production costs. Concurrent engineering is accomplished through integration of people, processes, standards and tools. It involves bringing together cross-functional and multi-disciplinary teams representing various functional areas. These professionals are brought together to discuss integrating issues of functional design, manufacturing, quality control, customer service, and so other relevant matters [14].

4.2 Systems Engineering

According to [15], "System engineering is an interdisciplinary approach and means to enable the realization of successful systems that satisfies the customer and stakeholder's needs in a high quality, trustworthy, cost efficient and schedule compliant manner throughout a system's entire life cycle. It focuses on defining customer needs and required functionality early in the development cycle, documenting requirements, and then proceeding with design synthesis and system validation while considering the complete problem." A system is defined as a construct or collection of different elements that together produce results not obtainable by the elements alone. The elements, or parts, can include people, hardware, software, facilities, policies, and documents; that is, all things required to produce systems-level results. The results are defined in terms of system level qualities, properties, characteristics, functions, behavior and performance [16], [17].

4.3 Lifecycle engineering

Lifecycle engineering refers to the consideration of the entire lifecycle of a system right from the point of conception of the idea to the end-of-life management of the system. The necessity for lifecycle engineering of a system arose from the realization that a new industrial culture has to be developed [18]. The new industrial culture has to drastically reduce the amount of wastes generated, reduce environmental damage and occupational health damages. Furthermore, the new industrial culture would need to increasingly reduce its dependence on non-renewable resources while adding to the overall effective utilization of invested resources. Sustainable industrial production is the new industrial culture that is environmentally conscious, economically sound and socio-culturally compatible to the community of its utilization/deployment. According to [19], Lifecycle engineering seeks to maximize a product's contribution to society while minimizing its cost to the manufacturer, the user, and the environment.

Lifecycle Engineering involves project definition, project planning, project plan implementation, monitoring process, evaluation and adjustment, closing process, and commissioning/celebration. Lifecycle engineering is based on determining all aspects of the design for the entire life of the system. All the stages of the system's lifecycle are evaluated at the beginning of the design process. Lifecycle engineering takes start-up cost and ongoing cost into account. It also takes into account the market trend, reliability, system replacement timeline, and the system's end-of-life management.

5. The Future of African Infrastructure System

Paraphrasing Lord Henley's perception on the characteristics of today's infrastructure in [20], the current nature of our transport, communication and energy and water supplies infrastructure makes them vulnerable to the effects of climate change. The reason for the vulnerability arose from two fronts. Firstly, they were not designed with the consideration of potential impacts of these facilities on climate change. Secondly, they were not also designed with the consideration of possible impacts of climate change on their functionality/operations and longevity in mind. To mitigate the potential risk and to build a sustainable future for Africa infrastructural development, indigenous engineers and stakeholders in African development would need to come up with or adopt a customized collaborative engineering approach to infrastructure development. Considering all the aforementioned engineering design approaches, customized collaborative systemic lifecycle engineering approach to infrastructure development is recommended/proposed. It is adjudged the best approach to infrastructure development that could be adapted to African culture to ensure the sustainability of African infrastructure development. Adopting the approach would facilitate resource use optimization, capacity building, reduced cost of ownership, and result in minimal contribution to climate change [21].

6. Implementation Process for Collaborative Lifecycle Engineering of Infrastructure System

Figure 1 is an illustration of the proposed collaborative lifecycle engineering approach that could be customized to facilitate the achievement of the desired sustainable African infrastructure development goals. It is an improvement on/adaptation of collaborative lifecycle design-based rural technology development process model developed by [22]. The application of the lifecycle design based technology development model starts with the assemblage of coalition partners to identify needs. The coalition partners may include product developers, policy makers, voluntary organizations, members of rural community groups and other stakeholders. Needs identification could be in the form of new infrastructure needs in the area. The need could also be for re-designing or retrofitting of an existing infrastructure to improve its performance or to address specific problem of the system. The next step involves making arrangement for how the needs will be addressed. This could be achieved through policy formulation, and the consideration and assemblage/incorporation of various design components and features. This step is followed by generating conceptual designs and applying various environmentally friendly design concepts (DFXs) on the conceptual designs. DFX refers to design for X where X could be material (DFMt), modularity (DFMd), assembly (DFMA), manufacturability (DFMf), disassembly (DFD), maintainability (DFS), use and reuse (DFUR), upgradeability (DFUG), remanufacturability (DFRm), recyclability (DFR), energy efficiency (DFEE), packaging (DFP), multi-lifecycle (DFML), multi-purpose (DFMP), or minimum residue (DFMR) and so on. Each conceptual design is then evaluated in turn for environmental friendliness by using E-LCA and by doing enabling infrastructural availability analysis (EIAA). They are also subjected to socio-cultural compatibility assessment (SCA), risk assessment (RA) and economic assessments (LCC). Environmental lifecycle assessment method (E-LCA) is used to screen each conceptual design in terms of the technology's potential resource requirements, emissions and their possible impacts at each stage of the technology's lifecycle and over its entire lifecycle. The potential impacts considered include global warming, ozone depletion, eutrophication, eco-toxicology and human toxicology. The best conceptual design is the one that has the lowest overall environmental impact potential. Lifecycle costing (LCC) is similar to the E-LCA, except that it is used to analyze the technology cost at each stage of its lifecycle. This helps us to account for the true and total cost of the technology. Social lifecycle assessment (S-LCA) is also a lifecycle analysis approach. It is used to determine the socio-cultural impacts of the technology at each stage of its lifecycle. Here we account for how it will affect the way of life of people who develop, use and live/work in the neighbourhood where the technology is developed and deployed. Risk Assessment: This is the determination of the potential impact of an individual risk associated with each stage of the lifecycle. The focus of this assessment is on the development and utilization of the technology. The assessment is done by measuring the likelihood of occurrence and the potential magnitude of the impact if it occurs. Cost–Benefit Analysis is used to compare the net present values expected to be derived from development or retrofitting of the infrastructure. Furthermore, before the evaluation of each option of the infrastructure system being considered for development, a minimum acceptable performance level has to be set for each of the evaluations described above. Any or all options that fail to reach such performance threshold is discarded. Those that survive the screening process are ranked according to their overall performance at each stage and throughout their entire lifecycle. For instance, assuming the conceptual design option A, C and G (out of options A, B, ... , K) satisfied the minimum requirements, the three of them will be ranked. A computer model or prototype of the highest ranking conceptual design option (say Option A) will be developed. The computer model will be simulated and evaluated using the lifecycle evaluation tools earlier discussed. If the results obtained from the computer simulation evaluation are acceptable, a scaled-down size of the infrastructure prototype would be to assess the real-life performance of the system. If /when the performance model is deemed satisfactory and there is confidence that it would adequately serve the need of the consumers, the appropriate community scaled size of the model then built, tested and chargehands from the community of deployment are then trained on the system. The skills development for the chargehands would involve training for complete know-how in the areas of operation and maintenance; managerial skills and knowledge for generating and managing technical change. The capacity building on the infrastructure system would be best achieved not only through the training of local technicians in each of the aforementioned areas, but also by continuous monitoring of the system on site and interactions with the operators of the system throughout the system's lifecycle. Doing so would enable the developer to step in whenever necessary. It would also afford the developer and the users to learn crucial lessons for further development of the system. Furthermore, it would prevent the breakdown or inadequate performance of the system. Consequently, it will assure the longevity of the system's lifecycle thereby ensuring the sustainability of the infrastructure. Thus, collaborative lifecycle engineering would deliver a consistent standard of living that can be maintained indefinitely for the present and future generations. This would be achieved by its helping to reduce negative impacts on human health and the environment, and by reducing costs to business and the entire community. Another way by which the approach would provide benefits to businesses and the community at large is through infrastructure system's lifecycle extension as it provides opportunity for continuous innovation and improvements [23].

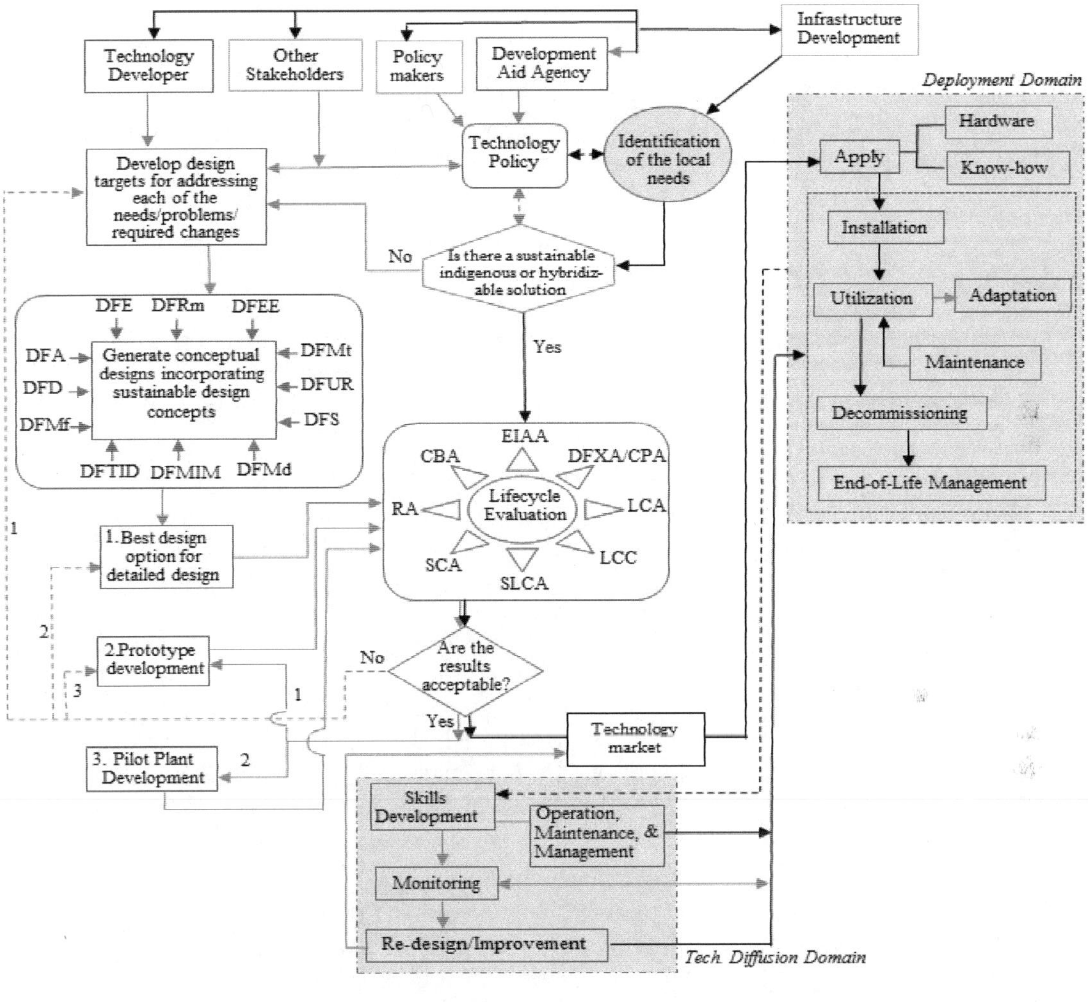

Figure 1 An Illustration of Collaborative Lifecycle Engineering Design Process for Sustainable Infrastructure Development

7. Conclusion

The current challenges being faced in the operational effectiveness of many infrastructures in Africa are attributable to a number of issues, some of which includes differences in climate and culture of the developers and the users of the technologies. The problem is expected to be exacerbated as the climate change due to incessant emission of greenhouse gases kicks in. There is therefore a need to prepare and implement design changes that would enable our infrastructures to adapt to the expected changes in our climatic conditions. African Engineers would need to come together and proactively meet the challenge of creating a climate resilient infrastructure system for Africa. This will reduce the risk of economic disruption to the continent and enable the opportunities from well-adapted infrastructure to be maximised. There is therefore a need for African engineers to rise up to our peculiar challenges by hybridizing the foreign infrastructure technology with our indigenous technology in designing, building and maintaining suitable infrastructure that are able to meet our current needs , consider future changes in our needs and that are able adapt to our culture and potential climate change in Africa. Collaborative lifecycle engineering is the engineering platform that would ensure sustainable development of African Infrastructure and assure their effective operation and utilization for our posterity.

References

[1] Pack, H. (1993). Productivity and industrial development in sub-Saharan Africa *World Development*, 21(1), 1-16

[2] UNCTAD (2015). Economic Development in Africa. Accessed online on 2nd May 2017 at http://unctad.org/en/PublicationsLibrary/aldcafrica2015_en.pdf.

[3] Perez, C. and Soete, L. (2017). Catching up in technology: entry barriers and windows of Opportunity. Accessed online on 6th May 2017 athttps://core.ac.uk/download/pdf/6786506.pdf

[4] Chileshe, N. (2016). Critical success factors for managing infrastructure projects in Africa: A critical review and lessons learned. In the Proceedings of International Conference on Infrastructure Development and Investment Strategies for Africa (DII-2016), Livingstone, Zambia. 31 August - 2 September 2016. Accessed online on 8th May 2017 at https://www.researchgate.net/publication/306079887_Critical_success_factors_for_managing_infrastructure_projects_in_Africa_A_critical_review_and_lessons_learned

[5] Chimbelu, C. (2016). Poor infrastructure is key obstacle to development in Africa. Accessed online on 8 May 2017 at http://www.dw.com/en/poor-infrastructure-is-key-obstacle-to-development-in-africa/a-15264436

[6] Dunmade, I.S (2002). Indicators of sustainability: assessing the suitability of a foreign technology for a developing economy. Technology in Society 24 (2002) 461–471

[7] Imaz, M. (2017). "Science and technology in the framework of the sustainable development goals". *World Journal of Science, Technology and Sustainable Development* (2042-5945),14 (1), 2

[8] Satapathy, S.K; Sangle, S. and Unnikrishnan, S. (2017). Role of government towards adoption of cleaner technologies for climate proactivity: A survey- based empirical study of Indian firms. World Journal of Science, Technology and Sustainable Development, 14 (1),29-49

[9] Dunmade, I.S (2012). The Use of Life Cycle Management Principles in Biosystems Engineering: A Pragmatic Approach to Solving Agri-industrial Sustainability Problems. Journal of Agricultural Science and Technology A 2 (2012) 357-362

[10] Gericke, K and Blessing, L. (2012). An Analysis of Design Process Models across Disciplines. the Proceedings of International Design Conference - Design 2012 held in Dubrovnik, Croatia From May 21-24, 2012. Accessed online on 2 May 2017 at https://www.google.ca/url?sa=t&rct=j&q=&esrc=s&source=web&cd=4&cad=rja&uact=8&ved=0a

hUKEwiQpKiF9NHTAhUO02MKHeJuBVYQFghCMAM&url=http%3A%2F%2Fm.designsociety.org%2Fdownload-publication%2F31984%2Fan_analysis_of_design_process_models_across_disciplines&usg=AFQjCNEq9FqchN2- TMeBI1fCLp6Wy1PLRQ&sig2=XsOj_wr9hQawkUttQ8ebvA

[11] Wynn, D. and Clarkson, J. (1984). Chapter 1 Models of designing. Accessed online on 13 February 2017 at http://www-edc.eng.cam.ac.uk/~nc266/RC15/Wynn+Clarkson.pdf

[12] Yassine, A and Braha, D. (2003). Complex concurrent engineering and the design structure Concurrent engineering: research and applications 11(3) p.165- 176. Accessed online on 7 May 2017 at http://citeseerx.ist.psu.edu/viewdoc/download?doi=10.1.1.481.8202&rep=rep1&type=pdf

[13] Hines, W. and Montgomery, D. (1990). Probability and Statistics in Engineering and Management Science,3rd Edition, Chapter 3, Wiley, New York. Accessed online on 6 May 2017 at http://www.me.nchu.edu.tw/lab/CIM/www/courses/Computer%20Integrated%20Manufacturing/Chapter3%20- Concurrent%20Engineering.pdf

[14] INCOSE (2017). What is Systems Engineering? Accessed online on 8th May 2017 at http://www.incose.org/AboutSE/WhatIsSE

[15] INCOSE (International Council On Systems Engineering) (2010). Systems Engineering Handbook: A Guide For System Life Cycle Processes and Activities. Accessed online on 8th May 2017 at https://teoriaorganizacional.files.wordpress.com/2014/05/se-handbook-2010-0201-nal-8- 5x11.pdf

[16] INCOSE (2012). Guide for the Application of Systems Engineering in Large Infrastructure Accessed online on 8th May 2017 at http://www.incose.org/docs/default-source/Working-Groups/infrastructure-wg- documents/guide_for_the_application_of_se_in_large-infrastructure-projects-2012-0625-to-approved-update-2013-0417.pdf?sfvrsn=10

[17] Alting, L. and Jorgensen,J. (1993). The Life Cycle Concept as a Basis for Sustainable Industrial Production. Annals of the CIRP, 42(1), 163-167

[18] Ishii, K. (1995). Life-Cycle Engineering Design. Transactions of the ASME, 117, 42-47

[19] ROE (Royal Academy of Engineering)(2011). Infrastructure, Engineering and Adaptation ClimateAdaptation ensuring services in an uncertain future. Accessed online on 24 April 2017 at http://www.raeng.org.uk/publications/reports/engineering-the-future

[20] Lu, B. (2011). "Systematic lifecycle design for sustainable product development". *Concurrent Engineering, research and applications (1063-293X)*, 19 (4), 307. **DOI:** 10.1177/1063293X11424513

[21] Dunmade, I.S (2010). Collaborative lifecycle design – A viable approach to sustainable rural technology development. International Journal of Technology Management & Sustainable Development, 9(2), 149- 158

[22] Dunmade, I.S (2001). Development of System Models for Industrial Processes Selection with regard to Product Lifecycle Extension (PLETS Models). Logos Verlag, Berlin, Germany. ISBN 978-3-89722-744-6